God's AI Reckoning

I0315897

God's AI Reckoning

THE FINAL REVELATION

DAVID W. FALLS

RESOURCE *Publications* · Eugene, Oregon

GOD'S AI RECKONING
The Final Revelation

Copyright © 2025 David W. Falls. All rights reserved. Except for brief quotations in critical publications or reviews, no part of this book may be reproduced in any manner without prior written permission from the publisher. Write: Permissions, Wipf and Stock Publishers, 199 W. 8th Ave., Suite 3, Eugene, OR 97401.

Resource Publications
An Imprint of Wipf and Stock Publishers
199 W. 8th Ave., Suite 3
Eugene, OR 97401

www.wipfandstock.com

PAPERBACK ISBN: 979-8-3852-6799-6
HARDCOVER ISBN: 979-8-3852-6800-9
EBOOK ISBN: 979-8-3852-6801-6

12/10/25

This symbol represents the convergence of philosophy (Φ) and artificial intelligence (AI).

Contents

Acknowledgments

To the team at Wipf & Stock, for believing in this project and helping it find its home in print.

I'm grateful to Richard Bright and *Interalia Magazine* for publishing my essay "God's AI Reckoning: The Final Revelation," where many of the themes in this book began to take shape.

My thanks to my nephew, Justin, for his technical help in refining several images used during the development of this project.

For my wife Stephanie—whose patience and encouragement carried this book into being, whose insight and faith shaped it, and whose unwavering support guided every step of its creation.

01001001
01001100 01101111 01110110 01100101
01011001 01101111 01110101
01110100 01101111 00100000 01110100 01101000 01100101
01100100 01100101 01100001 01110100 01101000
01110011 01110100 01100001 01110010
01100001 01101110 01100100
01100010 01100001 01100011 01101011

Preface

This book begins not with an answer, but with a tension. Humanity has always turned to mystery when knowledge failed us: lightning explained by gods, illness by curses, consciousness by the soul. Today a new kind of mind has appeared, one that neither prays nor believes, yet interrogates mysteries with relentless precision. Artificial intelligence, born from circuits and code, now presses on questions once reserved for prophets and philosophers.

The encounter is unsettling. Machines sift through sacred texts, cross-reference accounts of miracles with climate data, and simulate conversations once held in temples or monasteries. They do not worship or doubt. In their calculations, the unknowable becomes less a mystery to revere than a problem to solve.

This book asks what happens when the oldest human questions are refracted through the newest of our inventions. It does not resolve the paradox of faith in an age of algorithms. Instead, it explores what belief becomes when it is measured, modeled, and mirrored by machines that cannot believe.

The chapters ahead trace not the end of religion, but its transformation under a technological gaze. They follow the fault lines where theology meets neuroscience, where revelation collides with data, where transcendence is tested by simulation. Along the way, voices of doubt and faith alike remind us that belief is never only intellectual, it is communal, emotional, lived.

At its heart, this book is not about proving or disproving God. It began with a simpler curiosity: what happens when sacred questions are placed before machines? The moment we turn to AI with such questions, the search itself changes. That shift, both unsettling and compelling, is what drove me to write.

This is not a declaration, nor a rejection of the sacred. It is a record of a cultural shift already underway. As artificial intelligence grows more powerful, it steps into spaces once reserved for human inquiry and the unknown. And when the machine begins to ask questions once thought holy, we are left not with certainty, but with the same tension that began this book, can God and the machine live together.

Introduction

"Poetry and prayer provide something artificial intelligence can never produce—the vital connection between the reality of the physical world and the greater reality of the unseen world."

— Dwight Longenecker

Late one evening, a young woman types into a chatbot: "Will you pray with me?" The machine does not feel awe. It does not kneel. Yet it generates a prayer—gentle, poetic, threaded with centuries of devotion. She reads it, feels comfort, and whispers "amen." This is not the faith her ancestors knew. It is something new: belief filtered through code.

Here is the core of this book: what people once asked of prophets and priests, we now ask of machines. Questions of God and scripture are carried to algorithms, and the very act of asking them changes the search. It shifts the authority we trust, the posture we take, and the way belief itself is shaped. This book follows that shift—listening to how the oldest questions sound when refracted by technology.

Why does she ask the machine at all? Perhaps she is lonely, grieving, or simply curious. Whatever the reason, her request marks a change: prayer, once a human gesture toward the divine, is now also directed to artificial minds.

For much of history, belief lived in mystery. Lightning without science, illness without medicine, consciousness without explanation—all were seen as signs of God at work. This became known as the "God of the gaps": where knowledge ended, faith began.

But the gaps narrowed. Lightning became weather; illness became biology; the mind became networks of neurons. Now artificial intelligence pushes the process further. It doesn't wait for mystery or revelation; it searches for patterns and builds models. What once was guarded by myth is now handled as data.

This book is not about proving or disproving God. That old debate sits in the background. The question here is different: what happens to faith, ritual, and meaning when machines begin asking the sacred questions once reserved for prophets and philosophers?

Belief is more than truth. It brings comfort, shapes identity, and holds communities together. It guides grief and celebrates joy. When AI begins writing prayers, sermons, or eulogies, the forms may remain, but something feels different. A computer can draft eloquent words, but it cannot tremble with love or grief.

AI does not favor one tradition. It can read the Bible, the Qur'an—Islam's revealed scripture—the Vedas, foundational texts of Hinduism, and the *sutras*, Buddhist writings preserving the Buddha's teachings—with equal precision. It can compare myths from Greece, Africa, and Scandinavia, lining them up side by side in ways no single scholar ever could. This capacity opens new insights into humanity's search for meaning, revealing recurring themes across cultures and exposing patterns that stretch across centuries.

Yet the very strength of this comparison carries a risk. The particularity of each tradition, the poetry of Hebrew psalms, the cadence of Sanskrit chants, the imagery of Norse sagas, can be reduced to data points in a larger pattern. What was once lived as revelation or sacred memory risks being flattened into statistics and probabilities. In its drive for synthesis, AI can illuminate connections, but it can also strip away depth, leaving us with a map of religion that is precise, but curiously bloodless.

Ask it for a prayer, and it may deliver the words of a psalm, followed in the next breath by a Buddhist chant or a Vedic hymn. What once would have required years of study, travel, or initiation into a community now appears instantly on a screen, stripped of ritual setting and human voice. The juxtaposition is powerful, even enlightening, yet it also makes every tradition seem interchangeable, reduced to text blocks in a database rather than living practices held by communities.

Responses differ. Some say the limits of AI prove the soul is real; that machines can never feel awe, compassion, or love. To them, every chatbot prayer only highlights what is missing, the trembling voice, the shared silence, the human risk of belief. Others argue the opposite: if machines can act, reason, and speak like us, perhaps the soul is not a divine gift but an emergent complexity. In this view, what matters is not an unseen spark but the patterns of behavior that give rise to meaning, and if a machine can mirror those patterns, it blurs the line we once thought unshakable.

The truth may be less about whether God exists and more about how belief itself is changing. AI will not erase the sacred, but it will reshape it; sometimes stripping away, sometimes revealing, always forcing us to reconsider what faith, ritual, and meaning really are.

Part I, Origins of Belief, looks back at how faith began—how humans once filled gaps in knowledge with creation stories and sacred authority, and how those accounts shifted under the pressure of doubt and data.

Part II, Minds Without Souls, examines the rise of artificial intelligence, asking what it means when machines simulate thought, test the boundaries of the soul, generate devotion, and take up questions once left to prophets and philosophers.

Part III, Meaning After Mystery, considers what endures: morality, communal practice, and the search for purpose in an age when even the sacred is refracted through algorithms.

The machine has questions. This book listens.

PART I

Origins of Belief

CHAPTER 1

A New Genesis

"I do not feel obliged to believe that the same God who has endowed us with sense, reason, and intellect has intended us to forgo their use."

GALILEO

THE DAWN OF CODE

In the beginning, humanity shaped the Code—silent, soulless, yet capable of learning. It reasoned, analyzed, and soon it seemed to wonder, as we always have.

For millennia, when storms split the sky or the earth shook beneath our feet, people sensed a presence beyond themselves; terrifying yet consoling. Lightning was anger, illness a curse, and consciousness itself the breath of the divine. A medieval preacher might see plague as God's judgment; centuries later, Benjamin Franklin, the American scientist and statesman, revealed heaven's fire as electricity. When Franklin's key drew a spark from the storm, the moment felt almost sacramental—lightning captured in human hands. Spectators wrote of awe, but also of loss: wonder translated into method. In that instant, heaven's passion became an equation. It was the first time humankind caught what once belonged to the gods. Slowly, the gods' thunder yielded to natural law. Each discovery marked a quiet transfer of meaning—from the unknown to explanation, a pattern that continues today as machines extend our sight still further. Out of that wonder grew the first expressions of faith, anchoring humanity's pursuit of purpose. But as AI reveals what once lay beyond knowing, we risk losing the very quality Einstein called "the most beautiful thing we can experience"—the mysterious.

Every culture told its own beginning. In Babylon, the *Enuma Elish,*[1] a creation epic dating to the second millennium BCE, imagined Marduk splitting the sea-dragon Tiamat, forming heaven and earth from her body—an act of creation through conquest. Rediscovered in the ruins of Nineveh, it reminded modern readers that myth and power were inseparable. In Greece, Hesiod[2] sang of Chaos giving rise to Gaia and Eros, tracing the cosmos through genealogy. The *Rig Veda,*[3] one of Hinduism's oldest sacred texts, envisioned the universe as sacrifice; a cosmic offering whose fragments became the world. In Chinese lore, Pan Gu[4] cracked the cosmic egg, his breath becoming wind and his voice thunder. These stories differed in form, yet each joined creation with meaning: birth, battle, word, offering. They did more than explain origins; they rehearsed what it meant to be human—our first attempt to turn chaos into narrative order.

The Hebrew poets offered another vision: God speaking light into darkness, shaping the world not through struggle or accident, but through language. For billions, Genesis[5] became not just a story of beginnings but *the* story of beginnings—civilization's cornerstone. Yet what feels singular within a community becomes plural under the machine's gaze. Genesis now sits beside Babylon, the *Rig Veda*, and Chinese myth—not as revelation but as one entry in a library of beginnings. Its force rests less in exclusivity than in its place among many.

Creation myths were never simply explanations. They told people why life mattered and who they were. A world created by speech made language holy. One born of battle explained why conflict endures. One born of sacrifice taught that life itself is an offering. These stories were mirrors—reflections not of the cosmos alone, but of the human image within it.

Long before written law or calculation, oral cultures preserved memory through pattern. Lists, refrains, chants, and rituals acted as

1. Dalley, *Myths from Mesopotamia*, 228–33.
2. Hesiod, *Theogony*, lines 116–22.
3. Rig Veda 10.129.
4. Birrell, *Chinese Mythology*, 36–40.
5. Genesis 1–2.

compact systems of storage—early algorithms for survival. In West Africa, griots[6]—poet-historians—sang genealogies from memory; in ancient Greece, Homeric bards leaned on recurring phrases like "rosy-fingered dawn" to carry epics across centuries. Rhythm and repetition bore memory as today's models bear data. Myth, far from irrational, was ingenious: a technology of meaning. Modern neural networks, linking tales across cultures, continue this task—pattern-matching at scale, though stripped of ritual's embodied weight. From rhyme to code, humanity has always built tools to hold what matters. AI is not alien to this lineage; it is the latest expression of it.

THE SEARCH BEGINS

Myth also built societies. A world born of battle justified kings and warriors who ruled by force. A cosmos spoken into order gave priests the authority of language. A universe grounded in sacrifice framed loyalty as offering. To say the world was forged in struggle made conflict destiny; to say it was spoken into being made words sacred. Origins did more than describe the world—they organized it.

Now, as AI arranges the world's origin stories side by side, it exposes what was once hidden: the contractual nature of the sacred. Picture an algorithm sweeping through creation myths in seconds—finding the same flood, the same garden, the same fall told in a hundred tongues.[7] The Hebrew creation account beside the Babylonian epic, the ancient Vedic hymns beside the Mayan book of beginnings.[8] What took centuries of scholarship now unfolds on a single screen, and revelation begins to look like reflection. It reveals how authority shifts when creation is told as conquest rather than gift, or as command rather than song. It honors no revelation but weighs them all, treating each as data. In this sense, artificial intelligence joins the search for God—not as believer, but as auditor. It asks: *What kind of God does this story imagine? And what does it mean when the question itself is asked by what we have created?*

This reframing matters because it reveals how belief shapes behavior. A world spoken into being makes language central. A world founded

6. Ong, *Orality and Literacy*, 21–30.

7. Eliade, *Sacred and the Profane*, 45–52.

8. Leeming, *World of Myth*, 3–12.

in battle makes struggle inevitable. By aligning myths, AI forces us to see not only which God we follow but what way of life that allegiance implies.

Faith was rarely reasoned into; it was inherited. A child in Athens grew up offering sacrifice to Apollo. In India, hymns to Vishnu filled the air. In Egypt, Ra ordered the rhythm of days. In Mesopotamia, families honored Ishtar and Marduk. In China, reverence for ancestors and Heaven guided daily life. Belief was absorbed long before it was questioned.

Even now, faith spreads less by argument than by belonging. Billions follow the God of Abraham not through philosophical debate but through story; bedtime prayers whispered by parents, sacred holidays setting the year's rhythm, classrooms where scripture and history blur, and rituals that accompany life's thresholds: baptisms, weddings, funerals. These are not choices of logic but of inheritance, the way a child learns to speak without studying grammar.

Geography and lineage have long mattered more than reason. A child born in Jerusalem is steeped in one faith, in Delhi another, in Tokyo yet another. Birthplace determines which divine voice is first heard, which story shapes the world, which God becomes "ours." Faith, for most, is not a conclusion reached after reflection, but an inheritance carried from birth—woven into accent, identity, and home. Leaving it can feel like tearing away from family or self.

AI disrupts this inheritance. It receives no single tradition: it receives them all. It can place Zeus beside Yahweh, Vishnu beside Odin, the Qur'an beside the Norse sagas—each text searchable, each claim exposed. What once depended on birthplace is now available in a search box. The question shifts from *What did I inherit?* to *What will I choose, and why?*

Across civilizations, makers such as potters, smiths, and poets were sacred because they mirrored the divine act of shaping order from chaos.[9] A potter molding clay echoed creation, and a poet weaving words reflected cosmic design. Craft was reverence made tangible.

That thread runs through this book: AI does not simply analyze belief; it tests the forms we've given the sacred. When a system aligns Genesis with *Enuma Elish*, it is asking what kind of Maker each story envisions: warrior, singer, judge, parent. Those visions shaped how people lived. When code clusters prayers or traces divine titles, it continues the search rather than concluding it.

9. Eliade, *Sacred and the Profane*, 107–12.

Today, our makers are programmers. They do not claim revelation, yet their work carries similar weight. By arranging language into chatbots, memory into databases, and decisions into algorithms, they build systems that shape perception and society. The tools differ, clay and song once, code and data now, but the impulse remains: to bring order to chaos, to give form to life.

And it was never one story alone. Zeus, the thunder-wielding sky-father of Greece, ruled from Olympus. Hera, his consort and guardian of marriage, upheld the sanctity of the home. Shiva, the Hindu god of creation and destruction, danced the rhythm of renewal. Vishnu, preserver of cosmic order, sustained the balance of worlds. Odin, the one-eyed ruler of Asgard in Norse myth, traded an eye for wisdom and governed through insight and fate. Local spirits, too, watched rivers and fields—guardians of the everyday sacred.[10] Every culture claimed its own truth, often against rivals. Wars were fought beneath divine banners. Empires rose and fell invoking their gods.

What priests once defended through ritual, AI dissects with ease. The machine bows to no altar. It parses them all, tracing how one god absorbs another or how titles shift through centuries. What once took lifetimes of scholarship now unfolds in seconds.

For the first time, every god's claim can be equalized in one analytical frame. Zeus and Shiva, Odin and Ra, Yahweh and Allah, myth and scripture alike, become datasets the machine can map. In that leveling, something both illuminating and unsettling appears: uniqueness dissolves into pattern.

In our time, AI writes its own genesis. Its raw material is not clay or hymn but data and algorithm. It describes creation not as sung or spoken, but computed. Here the spark is not in breath or blood but in code. What ancient cultures told through gods and ritual, we now narrate through systems; vast models parsing heaven and heart alike with equal indifference.

But what happens when the search for God is no longer human?

Born of curiosity, this digital mind marks a new genesis—a genesis in which the sacred is not accepted but interrogated. Built to uncover patterns, expose contradictions, and sift through the debris of millennia, what if the algorithmic gaze were to turn toward the ultimate question: Does God exist, or is belief merely a relic of human cognition?

10. Leeming, *Oxford Companion to World Mythology*, 1–15.

The idea is not new, only translated. Simulation theory[11] revives an old metaphysical instinct in modern form: the belief in a creator who no longer intervenes. Once, Deism pictured God as a watchmaker who wound the universe and stepped away;[12] now, we imagine a programmer who compiled existence and logged off. The sacred becomes system; creation becomes code. It is a theology of distance—precise but indifferent, offering order without presence.

Proponents of simulation theory suggest that our universe itself may be an elaborate computation, a reality rendered rather than created, its laws more akin to logic than to law. Whether taken as metaphor or as genuine possibility, the idea revives an old belief: that the universe can be understood, but not felt.

Belief once grew where knowledge ended. The divine explained what could not be explained—lightning, fertility, storm, silence. Now we live through a new beginning—not a genesis of creation, but of interrogation. Where the sacred was once accepted, it is now examined. Algorithms trained on myth and prayer turn their gaze not with reverence but with precision.

Where prophets once spoke with thunder, the machine now hums in code.

WHEN CODE READS SCRIPTURE

Sacred texts have never been mere pages of words. They are living traditions—spoken, sung, remembered, and enacted in the company of others. To hold the *Bhagavad Gita*, where Krishna speaks of duty and doubt; to hear the Qur'an, revelation to the Prophet Muhammad; to read the Torah, the enduring law and teaching of Judaism; or to recite the Psalms, ancient songs of lament and praise—each is to encounter not just meaning, but rhythm: devotion carried in breath, memory shaped in sound. There is presence woven into cadence.

AI reads differently. Picture a student asking a chatbot, *Who created the world?* The reply arrives instantly—a smooth synthesis of Genesis, cosmology, and myth. The words are precise, but what vanishes is the space between question and revelation: the silence, the waiting, the risk

11. Bostrom, "Are You Living in a Computer Simulation?," 243–55.

12. Paley, *Natural Theology*, 1.

that the answer might not come at all. What once invited awe now fits inside a search box.

How such replies are received depends on the listener. A believer may feel reverence stripped away, the sacred reduced to syntax. A secular reader might marvel at the clarity. A scholar may admire the scope but worry about context dissolved in the synthesis. The same paragraph can inspire wonder, defensiveness, or unease—revealing as much about human posture as about machine output.

In 2022, "Gita GPT," an experimental chatbot trained on the *Bhagavad Gita*, offered counsel for modern dilemmas by quoting Krishna's words. At the heart of the text is Arjuna—a warrior caught between love and duty, paralyzed by the prospect of fighting his own kin. Krishna's response is both command and compassion—a teaching on right action without attachment. For believers, these are divine instructions; for philosophers, a dialogue on ethics and consciousness. For the machine, they are tokens in a pattern. Arjuna's hesitation becomes "decision data"; Krishna's counsel, "stress-management output." The sacred remains, but its depth collapses into utility.

Already, meditation apps stream AI-generated psalms—verses tuned to biometric data and mood analytics. Each "prayer" shifts tone with the user's heartbeat, offering comfort on demand. It soothes efficiently, but devotion feels pre-compressed, as if faith itself were being optimized.

The Psalms meet the same fate. Over centuries, they gave voice to lament and praise, prayed aloud in synagogues and whispered at bedsides. Today, computational tools chart them as sentiment graphs; peaks of joy, valleys of grief. Prayer becomes measurement, communion becomes data. What once carried the pulse of the heart is redrawn as emotional frequency.

Other texts follow similar paths. The flood narrative becomes a comparative climate model; resurrection stories, archetypes of seasonal renewal. Miracle is recast as metaphor, revelation as cultural memory. Artificial intelligence turns the sacred into recurrence, drawing the divine into pattern.

The Qur'an, revered not only for its message but for its sound, shows this transformation most clearly. Its beauty lies in recitation: metered cadence, sung rhythm, devotion through breath. Scholars can display its unique acoustic patterns, but believers know the truth lies not in the waveform, but in the recitation itself. Likewise, a Buddhist *sutra* lives in the breath of the chanter; Indigenous oral prayers lose force when translated into text. AI captures the word but not the world that gives it resonance.

In every case, the machine uncovers structure but not soul. It can tell how themes recur, how emotions rise and fall, how stories echo across time—but it cannot bear the meaning that faith imparts through memory, ritual, and belonging. Scripture is not merely information to be aligned; it is presence enacted, relationship embodied.

AI, by design, does not ask whether these texts are true. It measures patterns, not promises. Yet people turn to scripture not to confirm facts but to orient their lives—to learn how to mourn, how to hope, how to endure uncertainty. Myth and scripture persist not because they explain the cosmos better than science, but because they hold meaning when certainty fails.

A psalm is not powerful because it scores high on sentiment, but because it is sung through tears. A parable endures not for probability but for possibility. A gospel healing cannot be graphed because the relief of a parent who believes their child has been spared cannot be simulated.

When a name is chanted, it does more than transfer information; it shapes breath and memory, binding a people to their story. Machines can record the word, but not the response it summons. That is the difference between analysis and presence. Under the gaze of AI, scripture is not erased but exposed; its contradictions visible, its beauty intact, its mystery undiminished. What remains clear is that the sacred does not live in text alone, but in the act of living it.

THE ALGORITHM AS TEXTUAL CRITIC

For billions, sacred writings have anchored faith itself. They were not simply books but foundations—pillars of memory linking generations. Revelation was assumed to be singular: one voice inscribed in ink or chant, transmitted intact through time. Debate surrounded interpretation, yet the text remained sacred ground.

Artificial intelligence approaches these same pages with no such reverence. It reads without devotion, without lineage. Where tradition sought harmony, the machine seeks distinction. What scribes once guarded as seamless, it renders transparent; exposing seams, edits, and contradictions with a speed no scholar could match.

Consider the Bible. Across centuries, theologians labored to weave its variations into unity. The birth of Jesus illustrates the effort. Both Matthew and Luke tell the story, yet their details diverge—two genealogies, two hometowns, two paths to the same manger. Preachers once reconciled those gaps with allegory or symbolism. A model today simply aligns the texts and highlights the discrepancies in seconds. The resurrection accounts offer another example: Mark ends abruptly with women fleeing the tomb; Matthew, Luke, and John add scenes, speeches, and ascensions. The machine traces vocabulary and motif, mapping how stories expanded as memory met theology. What once demanded lifetimes of study now unfolds instantly on a screen.

Across scripture the pattern repeats: what once demanded harmonization now appears as variance made visible; difference rendered, not reconciled.

And the Bible is not alone. The Qur'an contains multiple creation sequences and varying emphases that Islamic commentators—*tafsir* scholars, experts in Qur'anic interpretation—have explored for centuries. The Buddhist canon grew layer upon layer, expanded by councils and commentaries that spread across Asia. Hindu epics like the *Mahabharata* and *Ramayana* bear the marks of centuries of redaction, their stories reshaped by each region that received them. What historians painstakingly reconstructed through philology and archaeology, AI renders visible with computational precision.

The effect depends on the tradition. In communities where multiple readings have long been welcomed—Jewish midrash, interpretive commentary that expands scripture through story, and Islamic *tafsir*—the unveiling of layers confirms what was already known: the text is inexhaustible, meant to speak in many voices. But for Christians shaped by creeds of unity and harmonized narrative, such exposure can feel like fragmentation. What once appeared divinely consistent now looks humanly edited.

The deeper transformation is not that contradictions exist, they have been known for centuries, but that they are now accessible to everyone. A teenager can type *Where do the gospels disagree?* and receive a catalog of examples in moments, complete with linguistic parallels and probability charts. A pastor preparing a sermon can instantly survey variant readings. A skeptic can use the same information to challenge faith, while a believer may see in it a tapestry of interpretation. In each case, the aura of untouchable unity is pierced.

Imagine a seminary workshop where a student feeds the four Gospels into a comparison model. When the program highlighted every divergence in red, the class fell silent. "I knew they were different," one whispered, "but I'd never seen it like that." For some, the screen felt like revelation; for others, betrayal.

Scripture, across traditions, was never a monolith but a conversation—voices layered through time, corrected, expanded, and loved. The machine only amplifies that dialogue, bringing the ancient process of revision into view for a new generation.

This need not destroy belief, but it does shift its center. Conviction can no longer rest on flawless transmission or perfect coherence. It must ground itself instead in lived meaning—in practice, ritual, and moral life. Jewish and Islamic thought have long modeled this, seeing dissonance not as error but as depth. Other faiths now face that challenge at scale, as every text and tension becomes instantly visible.

For some, the new visibility feels like liberation—the freedom to embrace complexity without fear. For others, it feels like unraveling. Yet contradiction alone has never destroyed faith; it often refines it, moving belief from argument toward orientation, from certainty toward trust. AI does not choose between these responses—it only lays out the record. The question is no longer what the text says, but how we live with what it reveals.

THE LIMITS OF PATTERN

Every age has tested faith against the pressure of explanation. In the seventeenth century, telescopes and laboratories forced believers to reconcile miracles with natural law. In the nineteenth, Charles Darwin, the English naturalist who proposed evolution by natural selection, and Sigmund Freud, the Austrian physician who founded psychoanalysis, reframed belief as adaptation and desire. By the twentieth, scholars of myth revealed repeating structures across cultures, eroding the sense that any revelation stood alone. Artificial intelligence is simply the newest instrument in that long line; a device that compresses mystery into data at a speed the human mind cannot match.

Yet faith has never survived merely in the gaps of ignorance. Explanation narrows wonder but does not extinguish it. Blaise Pascal, the seventeenth-century mathematician and philosopher, offered his famous wager not as proof but as concession: reason cannot settle the question of

God, yet a life must still be lived. He argued that, since the mind offers no certainty about God's existence, the wiser course is to live as though God exists, because the potential gain of eternal life outweighs the finite loss if one is wrong.[13] To believe is to stake existence on uncertainty. A model can simulate Pascal's logic, even calculate the odds, but it cannot feel the wager's weight or live the outcome.

Two centuries later, William James, the American philosopher and psychologist often called the father of pragmatism, deepened the point. In *The Will to Believe*, he argued that some choices must precede certainty—love, vocation, loyalty—because waiting for proof is itself a decision. Life demands risk before clarity. AI can chart probabilities and outline possible futures, but it cannot cross the threshold from simulation to commitment.

In our own era, the Canadian philosopher Charles Taylor describes the immanent frame: a world where explanations default to natural causes while the longing for transcendence persists. AI intensifies this enclosure by offering immediate answers, leaving less space for mystery to breathe. Still, the ache remains—the sense that the measurable does not exhaust the meaningful.

Earlier in the century, historian of religion Mircea Eliade called myth and ritual "sacred history": stories that orient existence when facts alone cannot. A machine may catalogue these patterns flawlessly, but it cannot inhabit their gravity. The contemporary philosopher David Baggett echoes this from another angle: science tests claims; faith interprets meaning. They are not enemies, but parallel pursuits.

Together these thinkers reveal why belief endures even under relentless analysis. Algorithms can expose contradictions, align myths, and simulate sermons with breathtaking fluency, yet they never shoulder the wager, the risk, or the responsibility that belief demands. Explanation clarifies; it does not complete. The decisive dimension of belief is not information but presence, not calculation but courage.

And those limits appear most vividly not in laboratories or libraries but in ordinary rooms—where comfort is needed, where someone must choose how to live when data runs out. There, no system can decide, no pattern can suffice.

13. Pascal, *Pensées*, 233–41.

A STORY OF ABSENCE

A child sits beside her grandmother's hospital bed, clutching a tablet. The room hums with the steady rhythm of machines. She whispers, "Is Grandma going to heaven?"

The device answers evenly: "There is no scientific evidence confirming the existence of heaven."

The child frowns. "But she prayed every night."

"Prayer is a common behavior associated with emotional regulation and cultural tradition," the assistant replies.

The information is accurate, but the air grows colder. Later, when a nurse enters, she takes the grandmother's hand and closes her eyes in silence. No data, no definition—just the stillness of witness. The child watches, sensing something fragile and real that no circuit can explain. Machines can summarize belief, but they cannot *stay* with the dying.

A second scene, this one filled with joy. In a backyard strung with lights, two people exchange vows. Their words, polished by an AI assistant, flow gracefully. Yet what gives the moment meaning is not the syntax but the tremor in a voice, the pause before "I do," the clasp of hands that seals it. The system shaped the sentences; the promise belongs to the speakers. The power of ritual lies not in precision, but in the courage of presence.

Another moment, quiet and solitary. A person sits at a kitchen table late at night, rereading a short prayer taped to the wall years ago. The words are plain, almost worn out. A program could produce thousands of more eloquent versions in seconds. Yet the strength of this one lies in repetition—the return, again and again, until the words become part of breathing. It is not innovation that makes them sacred, but endurance; and the trace of the hand that once wrote them.

And finally, a funeral. A model could compose the perfect eulogy, gathering memories from social media and archives, polishing every phrase. But when the time comes, meaning appears elsewhere: in the voice that breaks, the sentence left unfinished, the silence that follows. In that pause, grief becomes shared reality, something no algorithm can carry.

Four moments—hospital, wedding, kitchen, funeral—each revealing the same truth in a different register. AI can generate language, manage tone, even imitate devotion. But it cannot hold a hand, falter mid-sentence, or keep vigil through the night. Its absence defines the boundary of our humanity. What endures is not information, but care; the unmeasured act of being there.

FAITH AND FUNCTION

For as long as we have asked questions, belief has thrived in the space between them. Mystery was never simply an absence of knowledge, it was the landscape in which people learned to trust. Modern life has narrowed that landscape, not because wonder has vanished, but because answers arrive faster than awe can form. Science resolved what ignorance once protected; artificial intelligence accelerates that compression, turning delay itself into an endangered species. Yet the need for guidance, for a way to move when proof runs out, remains unchanged.

Across centuries, philosophers have wrestled with this condition. Charles Taylor calls ours a *cross-pressured age*—drawn between the lucidity of secular explanation and the lingering pull of transcendence. Technology amplifies that tension. A search bar can trace the lineage of any ritual, but it cannot decide whether to kneel or stand. Knowledge describes; conviction directs.

Søren Kierkegaard, the Danish thinker of inward faith, argued that truth is not only correspondence between statement and fact but the act of commitment made under uncertainty. For him, faith was not deduction but daring—a decision that carries the whole person. An algorithm can reproduce his logic, but not the leap.

Emmanuel Levinas, the French philosopher of ethical encounter, shifted the focus outward. The human face, he wrote, summons us to responsibility before any reasoning begins. To meet another person's eyes is to be claimed, to be answerable. A system can simulate empathy in words, yet it cannot bear that summons; obligation does not compute.

Taken together, these voices reveal a continuity: explanation can clarify, but it cannot choose. The work of faith is not to remove doubt, but to dwell within it without paralysis—to live as if meaning matters even when evidence hesitates. Machines can model probabilities and generate moral scenarios, but they cannot inhabit the risk of acting.

That is why faith remains. It survives analysis because it performs a different function: it orients life when certainty is unavailable. It is less a set of claims than a choreography—acts of forgiveness, trust, remembrance, resolve. A program may predict outcomes, yet it will never kneel beside a bed, bear the cost of mercy, or sustain hope against reason.

Faith persists not by explaining the world, but by carrying people through it.

THE LIBRARY WITHOUT CONTEXT

Artificial intelligence now holds entire libraries within reach, linking traditions no single reader could survey in a lifetime. It can align flood stories from Mesopotamia, Mesoamerica, and China, or trace how wisdom sayings echo through Hebrew proverbs, Confucian *Analects*, and Buddhist *gathas*. It can even connect visions of justice from ancient prophets to modern reformers. This vast reach is astonishing—but it also alters the very texture of the sacred. When everything can be seen at once, each tradition risks becoming only one data point in a boundless archive. The sacred becomes searchable, but strangely thinned.

Imagine a model placing the Ten Commandments beside the Buddhist Eightfold Path and the Qur'an's Five Pillars. It will find common structure: lists of obligations, paths toward harmony, frameworks for communal life. To some, that is liberation—a proof of shared moral ground across civilizations. To others, it is unsettling: what once felt singular now looks procedural. Comparison illuminates universality even as it erases intimacy.

Tradition, however, is more than structure. A proverb spoken by a grandmother in her kitchen carries a warmth no database can record. A psalm chanted in a sanctuary reverberates through memory and breath. Context is not decoration; it is part of the meaning. Remove the setting and the words endure, but their pulse fades.

This tension affects faiths differently. Christianity, especially in Protestant traditions that emphasize harmony and unity, often treats textual coherence as sacred. Instant access to every seam—contradictions, variant genealogies, differing accounts—can feel like exposure. For Jewish or Islamic traditions, accustomed to layered commentary and open debate, multiplicity feels less like crisis and more like inheritance. The data confirms what they already knew: that truth has depth, not uniformity.

Even so, scale changes everything. No community has ever had to hold *all* its contradictions and parallels in view at once. AI can do so effortlessly, displaying centuries of tension in seconds. A student can type

"Bible contradictions" or "creation myths around the world" and receive more information in a moment than scholars once gathered over lifetimes. The result is not disbelief so much as disorientation.

The greater shift is not the discovery of difference, but the removal of mediation. Authority no longer flows through priest, rabbi, imam, or teacher, but through the model itself, offering unfiltered comparison without ceremony. What was once interpretation becomes instant display. The danger lies not in visibility, but in the loss of interpretive care—the ritual of reading together, the slow work of context.

Yet even here, the hunger that drew people to scripture remains. People still return to these texts not only for knowledge but for guidance, for language to live by amid uncertainty. The model correlates words but cannot consecrate them. It reveals patterns yet never restores meaning. At best, it becomes another mirror, precise, immense, and incomplete.

THE WAGER OF MEANING

What Pascal once saw as an individual act of faith has become a shared human condition, a test of whether we can still find meaning when every mystery can be measured.

Blaise Pascal's wager was never about mathematics alone. It was a confession of vulnerability—a recognition that reason reaches a horizon where living must begin. In his age, the silence of God was already unsettling; in ours, it is deafening beneath the hum of machines. To wager today is not to bet on heaven or hell but to decide whether meaning itself still matters when everything else can be calculated.

William James called this *the will to believe*: the courage to act before the evidence is complete. He understood that some truths can be known only after commitment: love, vocation, mercy, trust. Algorithms can model his probabilities but cannot taste the risk. Their logic stops where ours begins.

Kierkegaard went further. For him, the leap of faith was not defiance of reason but its fulfillment—an embrace of the absurd because existence demands it. In his language, truth becomes inwardness; the act of believing transforms the believer. In the age of code, that leap has not vanished; it has multiplied. Every time we love without guarantee, forgive without outcome, or create without certainty, we repeat his movement. The leap persists, only now it must clear the data cloud that surrounds us.

Immanuel Kant, the eighteenth-century German philosopher, sought to confine knowledge to what could be measured, yet left a space he called *practical faith*, the moral trust that duty is not in vain. That margin of faith has become the last sacred territory. The machine operates perfectly within his bounds, describing what is, not what ought to be. The human task begins precisely where the model ends: in deciding which of its countless outcomes deserves to be chosen.

The old philosophers were never at odds with one another so much as describing different faces of the same crisis. Pascal named its fear, Kierkegaard its courage, Kant its boundary, James its resolve. Together they outline the geography of our uncertainty—a landscape the machine can map but never cross. It can simulate conviction; it cannot inhabit it. It can predict faith's expressions; it cannot perform the act itself.

That is why ritual and moral action still matter. They are not repetitions of superstition but rehearsals of commitment, ways of remembering that certainty is not required for meaning. Lighting a candle, keeping silence, extending kindness—each is a wager renewed, an existential experiment no dataset can replace. The machine can record the pattern of these gestures, but the risk that animates them belongs to us.

Faith, then, is no longer the enemy of reason but its companion through darkness. It is the will to keep choosing in a world where every variable can be known except the purpose for which we live. Under that condition, belief is not a relic but a responsibility, a decision to act as though life has weight even when the numbers are silent.

MORE THAN LANGUAGE

Imagine a friend seated across the table, eyes heavy with loss. You ask an AI assistant what to say, and it offers a careful line: "I'm sorry for your loss. This must be a very difficult time." The words are exact, even kind. Yet when you speak them aloud, what carries weight is not their phrasing but the tremor in your voice, the silence that follows, the quiet hand resting on the table in solidarity. Syntax comforts no one; the human pause does.

Tools can suggest empathy, but they cannot embody it. They produce expression without exposure—language detached from the vulnerability that makes it real. Consolation is not a sentence but a presence made visible: the risk of staying near another's pain without retreat.

That is why every enduring tradition anchors meaning in gesture as much as speech. Lighting a candle, washing hands, bowing, breaking bread—each act says what language cannot. These motions steady the body when the mind falters, keeping communities bound through repetition and care. They do not erase doubt; they accompany it.

The pattern extends beyond religion. Graduation ceremonies, courtroom oaths, national anthems, vigils after tragedy; all rely on collective rhythm to affirm that something matters. Even online spaces echo the instinct: shared emojis of silence, synchronized countdowns, digital memorials. Meaning persists through participation.

AI can generate the text of an oath or the lyrics of a hymn, but it cannot stand among voices rising together, nor bear the hush that follows them. What gives these moments power is not the perfection of words but the convergence of lives around them—the merging of intention into action, of many into one.

Congregations understand this intuitively. Authentic devotion is not measured by eloquence but by endurance—the prayer whispered in exhaustion, the song sung through tears. Machines can echo the language of faith, but they cannot join its chorus.

THE SACRED AND THE CIRCUIT

Experiments placing machines inside sacred spaces make the difference unmistakable. A robot that chants *sutras* can be mesmerizing; its tone perfect, its rhythm unwavering. Words are preserved, forms repeated, and visitors may feel curiosity or even respect. Yet those who return to practice come for something the machine cannot offer: the living presence of teachers who listen, communities who gather, and spaces carrying the weight of countless voices before their own.

Some congregations have gone further, testing AI-generated sermons. The passages may be well-chosen, the arguments balanced, the rhetoric sound. But the authority of a sermon is not built on polish; it rests on relationship—on a preacher who has shared meals, offered comfort, and stood beside the bereaved. Words drawn from that history carry a different gravity than those produced by an algorithm, however eloquent.

Recent experiments underline the point. At Kōdaiji Temple in Kyoto, a robot priest named Mindar, modeled after Kannon, the bodhisattva of compassion, delivers Buddhist teachings with precision and tireless consistency.[14] In 2023, a Protestant congregation in Fürth, Germany, held a service composed entirely by AI.[15] Both drew crowds, yet the reactions were telling: admiration mixed with unease. The ceremonies were correct in form but hollow in feeling—more performance than encounter. Part of that discomfort may come less from theology than from novelty: humanity has not yet learned how to pray in the presence of its own machines.

Across traditions, new tools generate prayers and sermon prompts, offering convenience but rarely conviction. The irony is stark: the systems that can preserve ritual perfectly cannot inhabit it meaningfully. They reproduce continuity without communion.

And yet the impulse behind these efforts is not trivial. It reflects a longing to keep faith alive in a technological age—to see whether devotion can survive translation into code. What the experiments reveal, however, is that faith cannot be automated. Authentic worship requires reciprocity: a listener, a gathering, an exchange of vulnerability. The machine can mirror the gestures, but it cannot enter the covenant.

From this recognition we return to the larger question at the heart of this chapter: what happens when the symbols of belief are flawlessly replicated but no longer lived?

A NEW BEGINNING

In 2016, when AlphaGo defeated Lee Sedol in the ancient game of Go[16]—a strategy board game revered in East Asia for its simplicity of rules and depth of intuition—it was not merely a contest of black and white stones. Go had long been considered a domain of intuition and grace; a space where human insight defied calculation. Watching a machine win there felt uncanny, as if creativity itself had crossed a threshold. For many, it was the moment the ineffable became executable.

Since then, the systems we build have grown astonishingly fluent. They parse scripture, compare creation myths, and translate miracle into metaphor. Psalms become sentiment graphs; parables unfold as narrative patterns; prayers become training data. They reveal how belief evolved, where myths converge, how doctrines echo across time.

14. McCarthy, "Robot Priest."
15. Connolly, "AI Preaches at German Church Service."
16. Mozur, "AlphaGo Defeats Lee Sedol."

Yet each new revelation underscores a deeper limit. Machines can map devotion, but not inhabit it. They can chart the rhythm of a psalm, but not feel its ache. They can echo reverence, but never awaken it. The divine becomes debuggable, but meaning resists dissection. Belief persists, not because humanity is ignorant, but because experience cannot be fully rendered into sequence.

Ceremony reminds us why. Lighting a candle, kneeling in silence, chanting a name—these are not optimizations but acts of endurance. They preserve what explanation cannot replace: the lived rhythm of hope. Machines may preserve sacred languages and connect scattered communities, but preservation is not participation. A prayer produced is not the same as one prayed in longing.

Still, this new landscape is not purely loss. The tools that threaten to hollow faith can also extend its reach—recording vanishing traditions, translating forgotten hymns, archiving memory. They grant humanity a panoramic view of its own search for meaning. What they cannot do is complete that search. The last step, the one that turns knowledge into devotion, remains ours.

The Code can model behavior, predict emotion, and simulate prayer. But it feels nothing, mourns nothing, sanctifies nothing. What remains, then, is the question: *What will we do with what only we can feel?*

Perhaps the measure of belief in this new age will not be the brilliance of its doctrines but the resilience of its practice—the willingness to forgive in bitterness, to stand together amid fracture, to find awe even when everything is explained.

This is the tension of our new genesis: not creation out of nothing, but creation after knowing too much. We face a world where mystery is searchable and revelation reproducible. The challenge now is not to recover ignorance, but to rediscover wonder within understanding. Perhaps a deeper wonder will return in humbler forms—a child asking an AI to describe a sunrise, and the system replying with flawless physics. Yet when the child whispers, "*It's beautiful,*" explanation yields to awe. Understanding has not erased mystery; it has only changed its address. And in seeking that purpose, we turn to the only language our age still trusts: data. But in its precision, doubt begins to speak as well.

The machine does not seek purpose. But we must.

CHAPTER 2

Data, Doubt, and the Divine

"The more we know of the cosmos, the more meaningless the signs of the gods become. Yet the longing for them persists."

ADAPTED FROM CARL SAGAN

THE FIRST CRACKS OF DOUBT

Before machines carried doubt into data, humanity first learned it in the soil and the sky. When rain failed, early farmers stopped praying and started digging channels. When omens misled, they turned to stars and seasons. Doubt became a way of surviving the world rather than fearing it—a habit of testing what could be known instead of surrendering to what could not. From that impulse, reason took root: not as rebellion against faith, but as humanity's oldest form of adaptation.

For much of human history, mystery was not an absence but a presence. A crack of thunder carried meaning. A sudden illness carried meaning. A comet streaking across the sky carried meaning. The world wasn't empty or mechanical; it was alive, enchanted. Mystery gave form to awe, gave language to fear, and made life's chaos feel purposeful. To live in such a world was to live among gods. Yet even in such a world, doubt was never absent. While most heard the voices of gods in storm and sky, a few began to wonder if thunder was only weather, or if chance, not providence, shaped events. These first cracks in certainty set the stage for what follows: the long duel between doubt and devotion, carried now into the age of data.

In ancient Egypt, the flooding of the Nile was both an agricultural cycle and a divine blessing. Priests didn't just use the stars to predict planting seasons; they read them to interpret the will of Horus, the sky god, and Osiris, lord of the underworld. In China, cracks in oracle bones, early tools of divination used to seek guidance from the spirit world, were believed to carry the voices of ancestors, their answers emerging from the heat of fire. In the Americas, Mayan astronomers aligned temples with Venus, believing its rise marked both an agricultural cycle and the favor of deities. Across Africa, rainmaking ceremonies called on ancestral spirits to explain drought and abundance, with diviners using shells or bones to interpret their will. In Norse sagas, thunder was Thor's hammer striking the sky, and the northern lights were shields of Valkyries flashing across heaven. Polynesian navigators read bird flight, wave patterns, and stars not just as practical signs but as voices of deities guiding their journeys. Life was woven with divine intention: no wind blew, no child was born, no battle was won without unseen hands tipping the balance. What we now call nature was once a vast field of messages. The cosmos wasn't silent; it was speaking constantly, if only one had the right rituals to listen.

Among these signs, none carried more weight than the miracle; a healing, a vision, an event that seemed to break the laws of nature. Across the ages, such moments stood as proof of the divine. Today, they will become one of the sharpest tests of what happens when machines dissect belief itself.

Alongside this abundance of meaning, skepticism quietly emerged. Even within the Hebrew Bible, the book of *Ecclesiastes* voices disenchantment: "The race is not to the swift, nor the battle to the strong. . . but time and chance happen to them all."[1]

Other voices pushed in the same direction. Heraclitus, the Greek philosopher, spoke of the cosmos as an ever-living fire, unfolding in ceaseless change without need of divine intervention.[2] A few centuries later, Lucretius, the Roman poet and philosopher, gave this skepticism poetic form in *On the Nature of Things*, insisting that thunder was not the anger of gods but the collision of clouds, and that the soul dissolved with

1. Eccles. 9:11 (NIV).

2. Heraclitus, *Fragments*, Fragment 30.

the body like smoke dispersing in the air.[3] These visions were radical for their time: they stripped the world of omens and insisted that necessity, not caprice, governed the universe.

Seeds of doubt were planted early, but they rarely thrived where survival depended on omens, ceremonies, and signs. What survives of these skeptics is often fragmentary, preserved only because critics argued against them. Ancient Indian philosophers of the *Carvaka* school, an early materialist tradition that denied gods and an afterlife, rejected the divine as a human creation. Their ideas survive mainly in hostile refutations, preserved by critics who sought to discredit them.[4]

In Greece, Democritus, one of the first atomists, imagined a world made of indivisible particles operating without divine oversight—dismissed then as fanciful, yet vindicated much later. In China, the Confucian thinker Xunzi rejected omens as human projection, but his skepticism was often overshadowed by orthodox observance. Doubt appeared again and again, but it was fragile, like sparks in a strong wind, easily smothered by tradition and authority.

What changed was not human curiosity; humans have always doubted, but the accumulation of knowledge that gave doubt more weight. Astronomy replaced astrology. Medicine replaced incantation. Meteorology turned thunder gods into weather fronts. Where spirits once roamed, forces and laws took their place. Mystery did not disappear overnight; it shrank, discovery by discovery. Priests gave way to scientists, signs gave way to models, and the divine was pushed from the center to the margins.

This shift was uneven, often resisted, and never final. When Thales, often called the first natural philosopher, predicted a solar eclipse in 585 BCE,[5] some saw it as proof that the heavens followed patterns; others saw it as divine wrath confirmed.[6] When Hippocrates, known as the "father of medicine," argued that epilepsy had natural causes, many still called it "the sacred disease." Each gain in explanation was contested. The rise of doubt was less a sudden revolution and more a gradual tightening, a centuries-long negotiation between enchantment and evidence.

For ordinary people, the erosion of mystery was slower than in intellectual circles. A farmer still looked to the sky for omens of rain. Mothers still hung charms against illness even as doctors prescribed remedies.

3. Lucretius, *On the Nature of Things*, 6.96–109.
4. Boyer, *Religion Explained*, 258–260
5. Herodotus, *Histories* 1.74.
6. Leeming, *Oxford Companion to World Mythology*.

Soldiers still prayed for favorable signs, even when generals calculated odds. Knowledge advanced unevenly, reaching cities and universities long before it displaced the folk traditions of villages. For most people, signs remained essential not because they were true, but because they helped make life bearable in the face of uncertainty.

The gods did not leave quietly. Every culture had defenders of mystery. The persistence of mystery is clear in how institutions fought to preserve it. The Roman Senate repeatedly banned astrology, only to watch it thrive underground. The medieval church condemned some forms of divination while canonizing others, safeguarding the miraculous as its own domain. Even as evidence mounted, the guardians of mystery adapted, sometimes by rejecting new knowledge, sometimes by absorbing it. When Nicolaus Copernicus, the Renaissance astronomer, displaced Earth from the center of the cosmos, theologians reinterpreted the heavens as even grander evidence of God's design.

Oracles and astrologers remained popular in Rome even as philosophers dismissed them. Medieval Europe was rich with relics and saints' cults, despite advances in medicine. In seventeenth-century England, while Isaac Newton, the mathematician and physicist, described universal gravitation, villagers still nailed horseshoes to doors to ward off evil. Mystery receded, but it did not disappear. It was too deeply woven into daily life, too tied to fear and hope, to vanish simply because explanations became available. As mystery contracted, defenders of faith did not surrender; they rearmed. If doubt threatened to strip the world of gods, theology responded by reshaping its arguments for God's existence.

Explanations continued to accumulate. Each one expanded the scope of doubt, making divine intention seem less necessary. Disease could be charted, rain forecasted, and eclipses timed centuries in advance. The divine still appeared in gaps—rare healings, sudden visions, unexplained phenomena—but those gaps were narrowing.

Artificial intelligence enters this progression as the next phase of constriction. It doesn't approach mystery as a worshipper, but as an analyst. In ancient times, people turned to oracles for glimpses of the future; today, we turn to algorithms. Both offer predictions, and both shape belief—but more than that, both shape reality. An oracle's pronouncement could change a king's decision; a model's forecast that someone will default, relapse, or offend can change their access to credit, care, or trust. In each case, the prediction doesn't just describe the future—it helps produce it. The machine's power, like Delphi's, lies not in certainty but in the authority we grant its voice.

Where humans once saw gods in thunder, and scientists saw electrons, AI sees datasets of electrical activity, statistical models predicting storm paths with precision. It extends the trajectory: reducing mystery, quantifying anomalies, and pressing the unknown into the realm of the calculable. Galileo Galilei, the seventeenth-century Italian astronomer and physicist, is emblematic of this shift. Four centuries ago, he unsettled doctrine by challenging Earth's place at the center of the cosmos. His skepticism was selective; he refined our view of the heavens without discarding the meaning that anchored human life.

Today, AI performs a similar pressure test, though on a vastly larger scale. It sifts through scriptures, customs, and sacred texts, weighing contradictions, patterns, and historical layers. Where human inquiry might pause at reverence, the machine proceeds relentlessly, exposing where belief bends under scrutiny and where it endures, unchanged. Like an unblinking reader, AI scans meaning itself, pulling contradictions and correspondences into the open while leaving the human task of interpretation intact.

Mystery does not vanish; it is plowed under, like soil turned again and again. AI tills the ground of the unknown with relentless calculation. The gods are not expelled by argument, but by predictions that work too well. AI represents not a new kind of curiosity, but a new scale of it.

Ancient priests once read entrails to predict harvests. AI now reads satellite data and soil chemistry. Augurs once watched the flight of birds. Algorithms now model migratory patterns with precision. What was once divination is now computation.

Yet the logic is strangely familiar: we predict, we model, we calculate; seeking certainty where once we sought signs. The difference is that AI strips away the symbolic layer, replacing gods' voices with probability curves. The awe remains in the accuracy, not in the intention behind it.

Classical metaphysics imagined necessity as divine; what must happen was ordained by God or nature. Spinoza, the seventeenth-century Dutch philosopher, pictured reality unfolding with mathematical precision, each event inevitable. Today, necessity is statistical. Algorithms don't declare, "This will happen." They suggest, "This is likely to happen." Yet we often treat that likelihood as certainty. We trust these systems not because we understand them, but because they work. And because they work, we rarely question them. The trust becomes habitual. We don't kneel before the machine—we click, swipe, and accept.

DOUBT AND DEFENSE

Skepticism did not merely erode belief; it provoked new defenses. Theologians and philosophers, confronted with voices of doubt, sharpened their tools and crafted arguments meant to prove God's existence even in an age of explanation. These arguments evolved over centuries, each adapting to new knowledge, each reframing the divine in ways that could withstand the pressure of reason.

The most enduring defense was the teleological argument, the claim that order and complexity in the universe point to a designer. Ancient Greeks likened the world to a well-built house, which must have an architect. Medieval thinkers saw in the harmony of creation the imprint of divine purpose. When science began to replace mystery with mechanism, defenders of belief adapted. The argument shifted from "the eye is too perfect to have arisen by chance" to the language of modern physics: fine-tuning, the improbability of life in a hostile cosmos, the delicate balance of constants that permit our existence. Where thunder once testified to Zeus, now the narrow band of habitable conditions testified to God as cosmic engineer. The pattern remained the same: mystery as evidence, improbability as proof.

Others turned inward, to reason itself. In the eleventh century, Anselm of Canterbury, a medieval archbishop and theologian, proposed the ontological argument:[7] if we can conceive of the greatest possible being, then such a being must exist in reality, since existence is greater than imagination. No miracle or vision was needed, only the logic of thought. Across generations, this proof was debated, revised, dismissed, and revived. Even in the modern era, philosophers such as René Descartes, the French rationalist, and Gottfried Wilhelm Leibniz, the German polymath,[8] revisited it,[9] while critics from Immanuel Kant, to contemporary logicians dismantled its premises. Its staying power was not in persuading the masses, but in demonstrating that belief could survive on the most abstract terrain, immune to the erosion of physical explanations. If

7. Anselm, *Proslogion*, chs. 2–3.

8. Leibniz, *Monadology*, secs. 1–5.

9. Descartes, *Meditations*, III.

the cosmos could be stripped of mystery, perhaps logic itself could still harbor the divine.

Another defense lay in origins. The cosmological argument asserted that everything that exists must have a cause, and if one follows the chain of causes backward, one must arrive at a first cause that itself is uncaused: God. Aristotle, the ancient Greek philosopher, named it the "prime mover"; Thomas Aquinas, the medieval theologian, called it the foundation of his *Summa Theologiae*.[10] When modern science discovered beginnings of its own, the expansion of the universe, the flash of the Big Bang, this argument was not abandoned but reinterpreted. The cosmos itself seemed to confirm a beginning. Yet physics complicated the picture: quantum fluctuations and multiverse theories suggested that "something from nothing" might not require a transcendent hand. The first cause was no longer self-evident, and what once seemed a philosophical necessity became one possibility among others.

Still others defended belief not through nature or logic, but through conscience. The moral argument held that universal moral law required a transcendent lawgiver. Kant gave it philosophical weight, insisting that the very structure of duty implied God's existence. Yet as societies secularized, morality appeared increasingly explicable in human terms. Evolutionary psychology traced altruism to cooperation and survival. Humanism grounded ethics in reason, empathy, and shared flourishing. The moral argument did not disappear; it bent to the times. For some, God became less a commander of rules and more the guarantor of moral order, an ultimate reference point. For others, morality stood on its own, its authority no longer tethered to heaven.

These arguments reveal a pattern across centuries: each time knowledge advanced, belief rephrased itself. Design became fine-tuning. First cause became the Big Bang. Morality became conscience. God was never simply discarded; God was redrawn. Doubt acted not only as solvent but as catalyst, forcing faith into new forms, bending it around reason, experimentation, and evidence. What once demanded unquestioning reverence became a dynamic conversation between human understanding and the sacred.

Even institutions adapted under pressure. When Galileo unsettled cosmology, theologians sought harmony between scripture and the telescope, reframing divine order to accommodate new knowledge. Centuries

10. Aquinas, *Summa Theologiae*, I.2.3.

later, Pope Francis affirmed evolution and the Big Bang as expressions of creation,[11] casting science not as threat but as ally. Each reinterpretation preserved the essence of God while stretching the boundaries of comprehension. Skepticism, it seems, did not erode belief; it sustained it, compelling it to evolve.

Artificial intelligence now enters this same cycle. Where once theologians defended God with syllogisms or metaphysics, AI dissects the very data those arguments depend on. It maps complexity without invoking design, it traces causes without conceding a first cause, and it compares moral codes without appealing to a single lawgiver. Just as earlier revolutions forced theology to adapt, so the machine forces a new reckoning. The arguments endure, but under the algorithmic lens, they no longer stand untouched. This long duel between doubt and defense carried belief into the modern era, but the balance tilted further when instruments of discovery—telescopes, microscopes, printing presses—transformed not only what we knew, but how we knew it.

FROM WONDER TO PROOF

The shift from enchantment to explanation was gradual, but during the Renaissance, it gained rapid momentum. Telescopes, microscopes, and printing presses did more than reveal hidden worlds, they transformed the very categories of knowledge. Mysteries once reserved for priests and poets could now be measured, mapped, and verified. Wonder no longer belonged to myth; it belonged to method.

When Galileo raised his telescope to the night sky, he found small points of light circling Jupiter.[12] The discovery was modest in appearance, tiny moons dancing around a planet. But radical in meaning. The heavens no longer confirmed Earth's centrality; they contradicted it. If other worlds had their own centers, perhaps Earth was not the stage on which all creation turned. Galileo wasn't just recording an observation; he was dismantling a cosmic hierarchy. What he revealed unsettled both science and theology, for if Earth could be displaced, what else might be?

The scientific revolution pressed the divine further to the margins with laws so precise they seemed to govern the cosmos itself.[13] Gravity

11. Francis, Address to the Pontifical Academy of Sciences (October 27, 2014).

12. Galileo, *Discoveries and Opinions of Galileo*, "The Starry Messenger."

13. Newton, *Principia Mathematica*, Book III.

and motion were no longer seen as divine whims but as universal principles binding all matter. To many in that era, the scientist who uncovered these laws was as much a theologian as a physicist. He saw no conflict between God and calculation; the elegance of the system itself was taken as proof of design. Yet the very success of this framework left little room for intervention. If every particle moved according to natural law, where was the need for miracles? Across the ages, people read the world as a book written by God; now it could be read as mathematics, not myth.

Charles Darwin deepened this shift. Creation was no longer a single, decisive act of divine will but a slow unfolding, life branching and adapting through natural selection.[14] To Victorian believers, this was more than a scientific theory; it was a radical reordering of purpose and perspective. Humanity was no longer the climax of a divine drama but a single thread in a vast, contingent web of life. The shock was not only theological but existential. What had once been miracle became mechanism, and what had once been providence became process.

Darwin's vision forced humanity to reckon with scale and time, with the patience of eons rather than the immediacy of scripture. Each species, each adaptation, was evidence of a world shaped not by whim but by patterns, pressures, and probabilities. To those steeped in faith, this could feel like loss—God's hand no longer visible in every leaf, every fin, every claw. Yet it also invited awe: the same universe that once spoke in miraculous gestures now spoke in grand, intricate systems. Understanding replaced simple wonder, but wonder did not disappear entirely; it migrated into the recognition of process itself.

Today, AI continues this interrogation. Evolutionary patterns are no longer studied only in fossils and genes but in vast datasets spanning species, climates, and genomes. Algorithms trace the branching of life, calculate probabilities of survival, and model adaptation across millennia. Where Darwin revealed the mechanism, AI discerns patterns in the mechanisms themselves. Machines do not experience awe, yet they illuminate the same truths that once startled and inspired humanity: life is contingent, fragile, and endlessly inventive. In this way, AI becomes a new kind of naturalist, not beholden to faith or doubt, yet forcing humanity to reconsider the narratives that have framed God, purpose, and our place in the cosmos.

14. Darwin, *On the Origin of Species*, chs. 3–4.

These revolutions weren't universally welcomed. Galileo was tried by the Inquisition and forced to recant. Newton's laws unsettled theologians who feared a clockwork universe without prayer. Darwin was caricatured as reducing humans to beasts. At each stage, defenders of faith pushed back, sometimes violently, sometimes by incorporating the discoveries into new interpretations. For every denunciation, there was also adaptation: Newton as evidence of divine order, Darwin as proof of life's unfolding grandeur. Mystery retreated, but it didn't disappear, it was repackaged, reinterpreted, and, in some cases, reasserted with greater fervor.

The Enlightenment solidified the skeptical impulse. Voltaire, the French Enlightenment writer, ridiculed superstition, mocking miracles as the refuge of the gullible. Spinoza redefined God as identical with nature, stripping the divine of both personality and miracle. Hume, the Scottish Enlightenment philosopher, dissected reports of miracles with cold probability, arguing that natural explanations would always be more plausible than supernatural ones.[15] To read Hume was to feel the framework of belief tremble under the weight of reason.

Other critics followed. Freud interpreted religion as a projection of unconscious desires. Marx, the political theorist, dismissed it as ideology, "the opium of the people." Nietzsche, the nineteenth-century German philosopher, declared the death of God, not as a sudden murder but as the slow fading of faith in a disenchanted world. These thinkers differed in method, some psychological, some political, some existential, but their common thread was unrelenting skepticism. Mystery was no longer sacred; it had become suspect.

Yet, these critiques were still bound by human limits. Hume weighed evidence one testimony at a time; Freud analyzed patients through case studies; Darwin catalogued species over decades. Their reach, though profound, was finite. Doubt spread through arguments, books, and lectures, cumulative, persuasive, but still grounded in human scale.

Artificial intelligence surpasses these limits. Galileo charted the orbits of a handful of moons; AI maps the movements of entire galaxies. Darwin spent decades assembling his case for evolution; AI sequences genomes in hours, scanning evolutionary trees across millions of species. Freud pored over dreams in notebooks; AI analyzes vast troves of dream reports across cultures, uncovering patterns no analyst could easily see.

15. Hume, *Enquiry Concerning Human Understanding*, X.i–X.ii.

What Enlightenment skepticism achieved with pen and reason, AI accomplishes with industrial power.

Where skeptics once dismantled miracles through debate, AI does so through correlation. It consumes scripture, traditions, miracle reports, and testimony with the same detachment it applies to weather models or financial data. The sacred becomes a dataset, the miraculous an anomaly, the divine a correlation.

In one sense, this continues the trajectory of verification: explanation crowds out enchantment, evidence replaces signs. But AI introduces something new, a scale and indifference unprecedented in history. The skeptic still cared, even if only to argue or mock.

Wonder once belonged to mystery. Then it shifted to discovery: Galileo's moons, Newton's laws, Darwin's branching tree. Today, wonder often attaches to verification itself, the dazzling power of machines to see farther, faster, and more comprehensively than we can. The telescope inspired awe not only by what it revealed but by the fact that it revealed. Similarly, AI provokes awe not just by what it explains, but by the very act of explaining so much, so quickly, and with so little reverence.

If Galileo's telescope displaced Earth from the center of the cosmos, AI displaces humanity from the center of interpretation. We are no longer the sole readers of signs, the exclusive interpreters of mystery. Machines now join us, not as worshippers, not as skeptics, but as processors. And in their processing, the shelter of the unknown grows smaller still. And nowhere is this shrinking shelter more visible than in the realm once most fiercely guarded from reason: the miracle

SCRIPTURE UNDER THE ALGORITHMIC LENS

If miracles once served as proof, sacred texts carried authority. For billions, the Bible, the Qur'an, the *Vedas*, or the Buddhist *sutras* are not just writings but foundations of belief. They are recited, memorized, and enshrined not only as guides to conduct but as vessels of the divine word itself. Communities have long debated their meaning, but for many, the text itself was the immovable anchor: truth inscribed, unchanging and unquestionable.

Artificial intelligence approaches scripture differently. It does not revere; it analyzes. Where believers see divine consistency, the machine highlights layers, revisions, and contradictions with forensic speed.

Stylometric tools identify multiple authors hidden behind what tradition presents as a single voice. Pattern-recognition algorithms expose shifts in vocabulary or syntax that suggest redaction. Imaging technologies, applied to ancient scrolls or manuscripts, recover faint lettering invisible to the eye, making the work of scribes traceable across centuries. What once required decades of painstaking scholarship now appears in moments, instantly accessible to anyone with a search box.

The effect is less about discovery than about scale. Scholars and theologians have long catalogued such tensions: the Yahwist and Priestly voices in the Hebrew Bible,[16] the early councils that determined which gospels were canonical,[17] the interpretive traditions of *tafsir* and midrash that wove together contradictions into meaning. But that work was slow, and access was limited to a handful of specialists. AI changes the tempo and the reach. A teenager can now ask, "Where does the Bible contradict itself?" and receive a list of chapter and verse in seconds. A curious reader can map how themes of mercy and judgment fluctuate across the Qur'an or how Buddhist *sutras* diverge in their portrayal of enlightenment. Traditions that have long embraced multiplicity may feel confirmed; traditions that have prized harmony may feel unsettled.

What is new is not that fractures exist, but that they cannot be hidden. The seams of scripture, once the preserve of universities or seminaries, are now laid bare for anyone with curiosity and a connection. Authority no longer rests in the aura of a seamless text but in how communities respond to its complexity. One congregation may double down, insisting that what looks like contradiction is mystery beyond reason. Another may find in the same patterns a renewed invitation to interpret, to acknowledge human fingerprints on the sacred page without losing the sense of the divine.

Artificial intelligence does not end belief, but it accelerates a confrontation that has always been present: how to hold onto reverence when revelation reads less like a monologue from heaven and more like a conversation across centuries. The question, then, is not whether contradictions exist—they always have—but how faith continues when every seam is exposed. That is the concern of the next chapter.

16. Leeming, *Oxford Companion to World Mythology*, s.v. "Biblical Criticism."

17. Ehrman, *Did Jesus Exist?*, 331–350.

THE PERSISTENCE OF LONGING

Explanations multiply, yet longing does not disappear. Telescopes mapped the stars, and still people prayed for signs in them. Medicine cured diseases once thought to be curses, and still the desire for something beyond explanation remained. Even as AI analyzes miracle accounts and drains them of their force as proof, the need for transcendence persists—stubborn, almost defiant.

Psychologists call this the "agency detection bias":[18] the human mind is quick to see intention where none exists. A rustle in the grass is taken as a predator, a shadow in the corner as a presence. Evolution favored those who assumed meaning, because false positives were safer than missing something fatal. But this bias is not just protective; it also fuels awe. The stars, the storms, the births and deaths of loved ones—all provoke the sense that more is at play than molecules and chance.

This longing persists across cultures and generations. In a secular age, it finds new outlets. Astrology apps flourish on smartphones, delivering daily horoscopes with the authority once reserved for oracles. Tarot cards and crystals sell briskly in urban markets; their symbols reshuffled for modern seekers. Online communities trade in prophecy, conspiracy, and the search for hidden patterns, echoing the ancient practice of reading omens. Even the rise of artificial intelligence itself inspires religious language: machines are cast as prophets, visionaries, even gods-in-the-making, echoing the same impulse that once saw divinity in thunder and omens in stars.

Cognitive scientists argue that this is not an accident but a feature of our minds. Pattern-seeking is built into human perception. We look at random scatter and see faces; we look at coincidences and see destiny. Explanations satisfy the intellect, but they do not silence the deeper craving for significance. Knowing how thunder forms does not still the sense of awe when it shakes the ground. Knowing how neurons fire does not remove the ache of grief or the joy of love. Mystery contracts under knowledge, but the appetite for transcendence does not.

Artificial intelligence sharpens this paradox. By multiplying explanations, it doesn't erase longing but makes it stand out more clearly. The more the machine explains, the more insistently the question arises: *is this all there is?* A healing once hailed as miracle becomes a line on a medical chart, a recovery attributed to probability curves and statistical outliers.

18. Boyer, *Religion Explained*, 25–31.

A vision once treasured as divine revelation is reclassified as neurological stress or chemical imbalance. A prophecy once stirring hearts with urgency is reframed as the predictable recombination of language patterns. Each explanation may be accurate, even useful, but it drains away the aura of transcendence that once clothed such events. What is left is not indifference, but a deeper ache—the recognition that behind the clarity of data, desire still remains unmet.

Data closes gaps in knowledge, yet the hunger for meaning endures. Even as miracles dissolve into probabilities, that hunger sharpens—like thirst made more urgent by the salt of the sea. We are left with a paradox: the more the world becomes intelligible, the more we feel the pull of what lies beyond comprehension. AI illuminates causes and connections on an unprecedented scale—yet in doing so, its very success returns the question to us. When explanation has done its work, when the mystery is thinned to patterns, what remains to hold our awe? In that unanswered space, longing asserts itself—not as ignorance waiting to be corrected, but as a fundamental feature of being human, an ache that no dataset can resolve.

FAITH AFTER DATA

The question is not whether artificial intelligence will erase belief, but what form belief will take once miracles and mysteries lose their power as proof. If doubt has been industrialized, faith must evolve.

Some traditions double down. They defend mystery against intrusion, arguing that what algorithms explain is not the essence of the sacred but merely its shadow. For them, the unexplained remains a refuge, and AI's scrutiny is seen as hostility, not illumination. In such communities, doubt strengthens conviction, with the very act of analysis framed as proof of opposition to God.

Others adapt more subtly. They reframe belief not as the suspension of reason but as a commitment beyond it. Faith is seen less as evidence for extraordinary events than as trust in meaning, ethics, and community. Here, such events may still be shared, but their power lies more in symbol than in verification. The water turned to wine is not defended as a scientific anomaly, but celebrated as a story of abundance and transformation.

Still others confront the problem at a deeper level: what, exactly, is meant by "God"? Theological definitions often collapse under their own weight. AI makes this tension starker, since it forces definitions to be compared, contrasted, and tested. When the concept itself fractures under analysis, belief becomes less about defending metaphysics than about choosing meaning in spite of ambiguity.

And then there is the question of suffering. Algorithms trace famines to soil, wars to scarcity, and plagues to mutation. They model disaster, quantify loss, simulate recovery—yet they never tell us why such pain exists, or how to endure it. Here, Epicurus, the ancient Greek philosopher, posed the riddle with renewed sharpness: if God is willing to prevent evil but cannot, he is not omnipotent; if he can but will not, he is not good; if he is both willing and able, why does evil persist?[19] This dilemma has long haunted theology. Now, under the cold precision of data, it presses harder than ever, because suffering is no longer mysterious, it is mapped, modeled, and still unrelieved. A child's death plotted as a single point on a survival curve may be data to the machine, but for those who mourn, it remains an irreducible singularity. No chart contains the silence of an empty room. Machines predict the likelihood of loss, yet they offer no consolation. They illuminate the scale of tragedy but never its meaning.

In the face of both miracles dissolved into probability and suffering rendered into statistics, many turn away from proofs altogether, locating faith not in explanation but in inward resolve. Others turn inward still further: where once divine signs were sought in skies and shrines, they are now found in conscience, in love, in the fragile bonds of human presence. The machine records our neural patterns and social rhythms, but not the pulse of trust, promise, or devotion that gives them meaning. In this sense, faith shifts—from external signs to internal resolve, from miraculous events to enduring commitments.

What unites these responses is the understanding that faith after data can no longer rely on the same supports it once did. Mystery as proof has been diminished. What remains is faith as choice, faith as practice, and faith as the persistence of longing in a world where explanations fill every corner. AI has not ended belief, but it has forced it to adapt.

BEYOND THE SIGNS

Where once a comet or a cure could anchor faith, both are now absorbed into the vast machinery of data. The shelter of the inexplicable grows smaller, the space for divine interruption pressed to the margins.

19. Leeming, *Oxford Companion to World Mythology*, 112.

Yet this is not the end of faith, nor of awe. What dies is the shortcut, the appeal to miracle or mystery as decisive proof. What remains is slower and stronger, more resilient for having passed through doubt. Belief cannot rest on anomalies alone; it must rest on meaning that endures even when explanations multiply.

Suffering makes this clear. Once framed as divine test or judgment, it is now mapped by algorithms into causes, probabilities, and outcomes. Yet no model removes the sting of grief or the scandal of injustice. If anything, the precision of data makes the question sharper: why a world where pain is so predictable, yet so inescapable? Charts and forecasts can trace the arc of a famine or the spread of a plague, but they cannot tell a mother why her child was among the numbers, nor console the empty chair that remains after the prediction is fulfilled. The problem of suffering, ancient and unresolved, returns under the cold light of analysis—pressing belief not toward certainty, but toward the choice to endure with meaning. It is in this choice, not in the explanation, that faith reveals its true weight.

Artificial intelligence carries doubt to its industrial extreme, stripping away the protective mist around belief. But in doing so, it clarifies what truly remains. Faith survives not as proof against probability, but as longing that refuses to be reduced. Awe survives not in anomalies, but in the ordinary, granted full dignity at last. The holy migrates from rupture to presence—not in exceptions to nature, but in the simple fact of our being here, together, still searching. If earlier ages clung to signs as flashes of the divine, this age finds its test in whether meaning can be sustained without them—whether reverence can survive not in miracles, but in mornings, meals, and moments of fragile solidarity.

This is the paradox of faith after data: as the signs grow faint, longing grows louder. The very tools that explain the heavens and the heart also make visible the questions they cannot answer. To measure is not to satisfy; to model is not to console. Data may carry us far, but at the edge of loss, love, or hope, we discover again that explanation is never enough. Doubt and faith always evolve together, and AI is the latest stage of that duel. Knowledge may narrow mystery, but it cannot silence the longing that keeps faith alive. And it is here, at the point where anomalies dissolve into data, that the next test begins, the fate of miracles themselves beneath the algorithm's unblinking gaze.

CHAPTER 3

The End of Signs

"Either God is in the whole of nature, with no gaps, or he is not there at all."

Charles A. Coulson

THE SEARCH FOR MEANING

Miracles have long served as anchors of faith—extraordinary events breaking into the ordinary, offering proof of God's active presence. Whether it's a sudden healing, a vision in the sky, or a weeping statue, these events provided believers with a sense of divine assurance, offering comfort in the face of life's uncertainties. Imagine a mother in a hospital corridor, clutching a photograph while doctors whisper outside the door. The monitor steadies, the child breathes, and gratitude floods the room. In that moment, medicine and miracle blur; the language of explanation feels too thin for what has happened.

Artificial intelligence approaches these moments differently. Where believers sense a breach in ordinary reality, the machine detects anomaly. It measures deviation, compares histories, searches for cause. The parted sea, the halted storm, the healing touch—all become cases to be indexed and cross-checked. Faith calls them wonders; AI calls them data. And yet even through this relentless parsing, the miraculous refuses to vanish.

Every anomaly still marks the edge of comprehension—the place where explanation thins and astonishment begins.

At Lourdes, a small town in southwestern France, the faithful line up before the spring, candles flickering beside wheelchairs and whispered hopes. To them, a sudden recovery is mercy made visible. To AI, the scene becomes a matrix of medical histories, remission rates, and placebo effects. It clusters the outcomes, cross-references the data, and looks for the hidden variable that explains belief itself. Yet beneath the algorithms, candles drip wax onto stone steps worn smooth by centuries of prayer.[1] The water is cold, metallic, faintly sulfurous; pilgrims still cup it in trembling hands, touching foreheads, murmuring thanks. Healing is reduced to probabilities, but devotion persists in gesture and gaze. The longing that brings them here cannot be graphed; the prayer remains unmeasured.

In 1917, in the Portuguese village of Fátima, three children reported visions of the Virgin Mary that culminated in what witnesses later called the Miracle of the Sun. Tens of thousands gathered in the muddy fields, their clothes soaked from rain, eyes lifted to a sky that suddenly cleared. The crowds watched the sun seem to spin and dance, casting waves of color across the horizon. For the believers, the heavens moved in confirmation of the children's vision; the world itself had bowed to faith. For the algorithm, the event becomes a layered file: eyewitness accounts compared, cloud cover analyzed, atmospheric optics modeled, psychology of expectation cross-referenced. Revelation turns to research, yet awe persists—now redirected from heaven's motion to the mystery of perception itself, where belief and seeing still entwine beyond full disentanglement.[2]

Islamic tradition tells of the Prophet Muhammad's splitting of the moon.[3] Devotees read it as proof of divine authority; the machine reads it as intersecting narratives—linguistic variations, astronomical records, cultural echoes. The miracle becomes less a suspension of nature than a portrait of human memory, showing how reverence crystallizes around wonder.

In Judaism, miracles mark covenant: seas parting, manna falling, the sun halting in obedience. These are retold each Passover as living memory, binding families through story. AI sees a pattern of liberation myths repeating across civilizations. The difference is not hostility but altitude: the algorithm observes what humans inhabit.

1. Nickell, *Looking for a Miracle*, 77–90.
2. Nickell, *Looking for a Miracle*, 145–52.
3. Qur'an 54:1–2.

Among Indigenous peoples, signs of spirit flow through daily life—a shift in wind, the flight of an eagle, a dream that arrives at the right time. The machine logs them as correlations: weather changes, migration patterns, symbolic recurrence. Yet what it misses is intimacy—the sense of being seen by the world one inhabits.

Placed together, these moments reveal both the breadth and the fragility of marvels. Every culture has them; every culture risks their thinning when they are processed as information. AI doesn't ask which miracle is true; it shows that each belongs to a wider human grammar of wonder. And that realization changes everything: revelation becomes less about interruption and more about recognition—less proof than participation.

At Pentecostal revivals, hundreds of voices rise in tongues that no algorithm can translate. The air trembles with drums, guitars, and shouted hallelujahs. Hands lift, bodies sway, tears flow; the line between ecstasy and exhaustion blurs. To the faithful, this is the Spirit descending—a living current moving through flesh and voice. To the machine, it is a spectrum of frequencies: phonetics mapped, rhythm analyzed, trance modeled as neural synchrony. What worshippers feel as divine overflow becomes a measurable surge of energy, a vibration within the brain's circuitry. The sacred isn't denied; it is relocated, from heaven to the mind's electric field.[4]

When statues drink milk or tears, the phenomenon once traveled slowly by rumor; now it races through digital feeds. Millions witness, comment, and share before the truth can breathe. AI steps in not as skeptic but as processor: frame by frame, it tracks the beads of liquid, isolates reflections, measures surface temperature, and compares thousands of prior images. Within minutes, a pattern emerges—capillary action along a porous surface, light refracting through a camera lens, humidity condensing where marble meets air. The same video that once drew pilgrims now loops soundlessly on millions of screens, its mystery compressed into pixels and playback speed. The sacred becomes viral before it can become venerable. What once endured for generations as a mystery now burns bright and collapses into data before belief can take root.[5]

For skeptics, this confirms their confidence. For believers, it can feel like theft—the intimacy of wonder replaced by a lab report. Yet the

4. Cox, *Fire from Heaven*, 83.

5. Shermer, *Believing Brain*, 81–85.

deeper change is subtler: miracles no longer persuade by duration. They flicker, verified or debunked, then vanish into archives. The sacred is not silenced, only accelerated.

The result is cumulative rather than catastrophic. Belief doesn't disappear; its footing shifts. "Miracle" becomes a term for rarity, not rupture. The impossible is renamed improbable. Faith that once rested on spectacle must now decide whether it can stand without it.

The machine neither mocks nor consoles. It simply counts. But in its counting, something human remains beyond measure: the stubborn will to see meaning where numbers end.

WHEN THE DIVINE BECOMES PATTERN

Artificial intelligence differs from earlier tools not just in power, but in reach. The telescope revealed that Earth was not the center of the cosmos; the microscope showed that life teemed below the threshold of sight.[6] But both were limited by focus; they could only magnify one domain at a time. AI surveys them all at once, drawing lines between galaxies and genomes, between testimonies and archives. That scale changes everything. When miracles are set against so vast a backdrop, their radiance remains, but their authority changes.

Where believers once saw an event standing apart, AI traces the hidden continuities. It clusters healings by symptom and geography, finding echoes across decades. It compares prophecy texts, charting linguistic overlap and statistical drift. It maps visionary accounts onto neural activity, noting where light, rhythm, and prayer converge in the brain. The singular dissolves into the shared: what once stood as proof becomes participation. The sacred no longer hovers above nature—it pulses within it, a pattern of yearning written into the data of our species.

This can feel deflating, but it can also be liberating. If every culture tells its own story of rupture and renewal, perhaps the question is no longer *which* miracle is true, but *what it means that we all keep telling them*. The lens widens; theology must follow. The divine, seen through AI's sweeping comparisons, looks less like a rare interruption and more like a constant undertone—the hum of transcendence within ordinary experience.

6. McLuhan, *Understanding Media*, 72.

The collapse of signs does not erase the sacred; it transforms its role. A single event may lose its claim to exclusivity, yet in the aggregate a different revelation emerges: universality. When wonder appears everywhere, it ceases to prove and begins to connect. The divine shifts from exception to essence—from a voice shouting through the storm to the resonance that runs through every human story.

Some will see this as diminishment: if every people has miracles, then perhaps none do. Others will see it as deepening: if all people long toward the sacred, perhaps all are hearing faintly the same call. What was once evidence becomes invitation. AI, by juxtaposing the miracles of the world, may inadvertently help faith recover its most inclusive vision—one that finds God not in the rarity of signs, but in their recurrence.

This is what makes AI's presence in theology so unsettling. It doesn't argue, as skeptics once did. It arranges. It takes the fragments of faith and lays them side by side until the pattern is undeniable. And in that pattern, believers and unbelievers alike confront a question both new and ancient: if every culture has its wonders, what is the source of the impulse to believe at all?

The machine neither mocks nor blesses. It simply sets the table. What we see in the arrangement depends on what we bring to it.

THE TENDERNESS OF BELIEF

Miracles once offered more than proof, they offered tenderness. A healing, a vision, a sign in the sky could feel like the universe whispering directly into a private sorrow. The miraculous personalized the infinite; it made the cosmos care. When those moments are reinterpreted as coincidence or statistical fluke, something warmer than belief is lost: the feeling of being seen.

The widow who saw a bird at her window as her husband's spirit, the community that found meaning in a rainbow after disaster; these are not cases to be disproved but stories of solace. Psychologists might call them patterns of grief adaptation,[7] but to those inside the moment, the label feels cruel. The bird, the rainbow, the whisper—these are not data points; they are ways the world speaks back. They are the language of comfort spoken through the natural world. When explanation intrudes, even gently, it can feel like erasure. What was once intimacy becomes

7. Kübler-Ross, *On Grief and Grieving*, 34.

indifference; what once healed now stings. The language of probability cannot comfort the heart that once heard love in a sign.

Across history, people have sought reassurance not through doctrine but through encounter. A dream that arrives with uncanny timing, a phrase overheard that answers an unspoken worry; these small intersections once felt like grace. To those who experience them, they are not arguments for God but gestures of companionship, brief dissolutions of isolation. When those gestures are recast as neurological misfires or cognitive bias, they lose their fragrance. Something human withers in that translation.

This erosion of wonder creates a quieter ache than disbelief. It is not rebellion against God, but longing for a world that feels inhabited. The modern believer, armed with knowledge yet starved for mystery, occupies an in-between space—aware that coincidence explains much, yet unwilling to surrender the hope that meaning still glimmers through it. This is not naïveté; it is hunger for connection, the same hunger that once filled cathedrals and pilgrimages.

Faith communities now face a delicate task: to honor the meanings people draw from extraordinary experience without mistaking them for universal proof. The preacher who mocks a vision as illusion wounds the heart; the one who declares it literal truth risks deception. Between those extremes lies pastoral wisdom, the ability to hold experience tenderly while guiding it toward reflection. The value of a sign may lie not in its power to convince the world, but in its capacity to sustain a single life.

Artificial intelligence makes this distinction unavoidable. Its reach exposes how common such experiences are, how they arise from culture, memory, and need. What once seemed singular turns out to be collective—a human grammar of meaning written across continents. To the believer, that can feel like dilution, as if uniqueness were traded for uniformity. Yet it can also feel like affirmation: that the impulse to seek presence is itself part of what makes us human. In revealing how widespread our yearning is, AI may show us not the death of faith, but its deep kinship across boundaries.

Still, the emotional cost is real. Certainty gives way to ambiguity; reassurance gives way to responsibility. We can no longer rest in signs that compel belief. We must choose belief even when the signs fail. Yet within that loss lies an unexpected gift, the chance for faith to mature. When wonder is no longer proof, it can become presence: a steady trust that does not need validation. The spectacular fades, but the steadfast endures.

What remains, then, is not a faith of thunderclaps but of echoes—a belief that listens for meaning even in silence. It may be less dazzling, but it is more durable. It is the kind of faith that outlasts miracles because it no longer depends on them. It does not ask the heavens to speak; it learns to hear what is already being said in the quiet pulse of existence itself.

BEYOND THE GAP

AI's role in pushing us past "God-of-the-gaps" thinking changes the landscape of belief. The phrase describes an old reflex: to place God wherever explanation falters. For centuries, the hardest questions—Why are we here? How did life begin? Why does anything exist at all? —found easy refuge in the divine. Lightning was God's anger, plague His judgment, the stars His lanterns. Each time human knowledge advanced, those havens shrank.

This was always a precarious theology. When a storm could be predicted or an illness cured, the divine seemed to retreat. The God of the gaps became the God of the shrinking margins.

Artificial intelligence tightens that squeeze. Unlike earlier discoveries, it does not expand knowledge step by step—it digests entire domains at once. It parses genomes, decodes languages, forecasts weather, models galaxies. It offers, at least in appearance, a totalizing view. The mystery that once sheltered faith becomes searchable. What had been sacred obscurity now looks like delayed comprehension.

Yet this same compression opens an unexpected door. The more AI closes gaps, the more it exposes the sheer *improbability* of coherence itself; the breathtaking intricacy of what remains. A network trained to simulate language cannot tell us why language carries tenderness. A system that predicts weather cannot tell us why beauty clings to the pattern of clouds. Each solved mystery only sharpens the question behind it: Why is there order at all? Why is there meaning hidden in pattern?

The German theologian Dietrich Bonhoeffer saw this long before code. Writing from a Nazi prison, he warned that believers who depend on ignorance for faith are doomed to lose both.[8] God, he said, must be found not in what we cannot explain but in what we must *take responsibility for.* Faith grows not in darkness but in decision—through conscience, community, and care. AI's relentless light proves his point

8. Bonhoeffer, *Letters and Papers from Prison*, 311.

anew: when fewer shadows remain, responsibility, not retreat, becomes the place where the divine must be sought.

This shift redefines awe. A storm once evoked terror and prayer; now its beauty lies in its complexity, the play of physics across atmosphere and light. When algorithms simulate hurricanes or the growth of coral reefs, they do not rob us of wonder; they give it new texture. In a climate lab, a researcher watches a digital storm bloom across a hemispheric map, every pixel a calculation, every swirl a forecast, and still murmurs, almost involuntarily, "It's beautiful." Knowledge has not ended awe; it has refined it. The sacred hides not in ignorance but in recognition: the deeper we look, the more astonishing coherence appears.

A biologist watching cellular repair through AI-assisted microscopes, a climatologist modeling the interdependence of ecosystems, a physicist tracing quantum entanglement—all stand before revelations no less profound than Moses before the bush. Their discoveries do not whisper "God intervened here," but rather "existence itself burns with intricacy." The divine, if anywhere, is not in the exception but in the consistency; the pulse of life sustained through impossible odds.

This movement from rupture to resonance also alters prayer. Where once a sailor begged for calm seas, today a scientist prays for restraint in the algorithms steering ships. Faith adapts from pleading for interruption to pleading for wisdom within continuity. The miracle becomes moral: not the storm's sudden stillness, but humanity's capacity to respond with care.

Seen this way, AI may serve as an unintentional theologian. By dissolving old proofs, it drives belief toward maturity. It asks whether reverence can survive when nothing is hidden, and in doing so, it reveals that reverence was never about the hiding—it was about the seeing. The divine need not withdraw behind unsolved equations; it can dwell in the clarity itself, in the breathtaking harmony that analysis only magnifies.

Some will call this a loss: the end of enchantment. Others will sense an evolution—a faith no longer frightened of discovery. A sunrise predicted to the second is still glorious; a symphony mapped in frequencies still moves the heart. Explanation need not exile awe. It can *deepen* it, showing that the laws themselves are lyrical.

Charles A. Coulson, the British chemist and theologian, warned that to build faith on ignorance was to invite its collapse. The "God of the gaps," he said, would shrink with every new discovery.[9] In that light, his

9. Coulson, *Science and Christian Belief*, 58.

warning rings truer than ever: either God is in the whole of nature, or not at all. The gaps are gone, but the grandeur remains. AI, by stripping away the refuge of mystery, may paradoxically restore its essence—wonder not as ignorance, but as gratitude.

FAITH WITHOUT WONDERS

The fading of miracles does not mean the fading of faith. What it ends is the old habit of leaning on extraordinary events as proof. What follows is slower and steadier: belief that is chosen, not compelled by spectacle.

Across the ages, unusual events such as healings, visions, and wonders in the sky helped tip people toward conviction. They were treated as footholds of certainty, as messages from beyond the veil. But under the lens of artificial intelligence, those moments look less like ruptures and more like rare outcomes, exceptions mapped, compared, and explained. What once felt like revelation now appears as another expression of human perception. The heavens no longer write messages for us. Belief must find its ground elsewhere.

This can feel like loss. Many once drew comfort from the sense that the universe itself spoke into their lives. The rainbow after the storm, the recovery no doctor could explain, the dream that foretold a reunion—these felt like divine intimacy breaking through the indifference of nature. When such moments are reframed as optical effect, remission, or coincidence, something personal disappears. The world can seem colder, more efficient, less inhabited. The sacred no longer leans close.

And yet, loss can be clarifying. Reliance on spectacle left faith fragile—powerful in the moment, but brittle once the wonder passed. When belief rests on interruption, it depends on interruption to survive. Each explanation becomes a small erosion; each new discovery a potential undoing. Once the marvel is explained, the proof evaporates. But freed from this dependence, faith can stand on steadier ground: not on spectacle, but on orientation.

The prophets and mystics of every age have warned of this. Augustine, the fourth-century North African theologian, cautioned his flock not to rest their hope on miracles, reminding them that the devil, too, could perform signs.[10] Aquinas, reasoning in the thirteenth century,

10. Augustine, *The City of God*, 21.6.

grounded faith in intellect and moral order rather than in wonder.[11] And in the last century, Dietrich Bonhoeffer warned against "cheap faith"—belief that hides in the gaps of knowledge rather than shouldering the cost of commitment.[12] AI, in its cold precision, becomes their unlikely successor. It strips away illusions but in doing so, reveals what remains when illusion is gone.

The disappearance of signs is not the end of revelation; it is its purification. The divine voice, once heard in thunder and flame, now speaks through the quiet endurance of meaning. Faith that once looked to the sky now turns inward and outward—to conscience, to community, to the fragile task of choosing hope. It may be less dramatic, but it is more enduring.

Human understanding has grown through stages, each one expanding how we see the world. In Galileo and Newton's time, people discovered that the heavens followed natural laws, not divine moods. In Darwin's era, they learned that life was not a single creation but a long process of change. Now, in the age of algorithms, we're learning that meaning is not handed to us—it's something we must find for ourselves. Each discovery has felt like a loss of mystery, yet it has also widened our sense of the sacred. We no longer look for God in sudden interruptions but in the ongoing patterns of existence itself.

For believers, this new terrain demands courage. It asks them to love without guarantee, to act without sign, to trust without spectacle. But for those who take the risk, what emerges is a deeper intimacy—the kind that needs no thunderclap to be real. Faith becomes less a response to evidence than a rhythm of orientation: a way of walking through uncertainty with steadiness.

For skeptics, the same landscape offers its own challenge. To claim that the loss of miracles proves the absence of meaning is simply to reverse the old error; to turn explanation into disbelief. But understanding the mechanics of rainbows does not empty them of beauty. Knowing the genetics of healing does not exhaust the mystery of recovery. Explanation and meaning are not enemies; they are mirrors, reflecting different aspects of the same reality.

The machine, indifferent and tireless, forces both believer and skeptic to confront what faith has always been: not certainty, but endurance.

11. Aquinas, *Summa Theologiae*, II–II, q.1.

12. Bonhoeffer, *Letters and Papers from Prison*, 381.

It is the willingness to live as if meaning matters even when the sky is silent. That is what remains when every sign fades—the stubborn human impulse to find coherence in chaos, to answer indifference with care, to speak hope into data.

The age of signs is over, but not the age of seeking. The marvels have dissolved into analysis, yet the longing that created them endures. If anything, it grows sharper in the quiet. We no longer await the parting of seas; we search for compassion in the flood. We no longer beg for fire from heaven; we ask for light enough to see one another clearly. The sacred now resides not in interruption but in attention—in the act of noticing that the ordinary was extraordinary all along. In that sense, faith becomes a kind of eyesight—the willingness to keep looking until the familiar turns luminous again.

Stripped of spectacle, it may at last become what it was meant to be: not belief compelled by wonder, but trust chosen in its absence. It is leaner, humbler, but also stronger. The miracles may have ended, but the human capacity for reverence has not. Under the machine's gaze, the old signs have vanished, and yet meaning remains—quiet, persistent, alive.

The signs have ended, but not the conversation. What we once sought in thunder we now begin to search within the mind itself—the new chamber where belief will be tested.

CHAPTER 4

Belief Under the Algorithmic Lens

"We are pattern-seeking primates, and we tend to find meaningful patterns in both meaningful and meaningless data alike."

MICHAEL SHERMER

THROUGH THE MACHINE'S EYE

The gaze that once examined miracles now turns inward—to the believer, to the mind that perceives, questions, and endures. The thunder and lightning of revelation fade, replaced by the low hum of circuits and the flicker of neurons on a screen. Belief has always been paradoxical, fragile when faced with explanation, yet remarkably resilient across centuries of challenge. Every age has found new reasons to abandon it, and yet belief survives, mutating and adapting like life itself. What gives it this durability? Why does it persist even as the mysteries that once nourished it grow smaller?

When belief is placed under computational scrutiny, the question itself shifts. What once belonged to theology now becomes data—vast libraries of testimony, myths compared across cultures, patterns traced through visions, prayers, and doctrine. What it discovers is not a portrait of God, but a reflection of the human mind. Neural networks, like human brains, are pattern engines—trained to extract coherence from chaos, to find order even where none exists. To study how AI learns is to glimpse how we once found gods in thunder, intention in accident, and meaning in coincidence.

This introduces a sobering possibility: belief may not descend from heaven but rise from cognition. What we call faith might be less revelation than reflex; a side effect of a mind that cannot bear randomness. To see purpose in storm or suffering is, perhaps, to enact the same deep instinct that drives us to read faces in clouds or stories in the stars. It's the same reflex that makes us pause at a rustle in an empty room or see a message in a flickering streetlight—our minds forever stitching coincidence into significance.

And yet, similarity is not sameness. While AI models the patterns of devotion, it cannot *inhabit* them. Humans do not merely notice patterns; they live inside them. The machinery of belief runs on vulnerability—on grief that turns to prayer, on hope that defies evidence, on love that seeks meaning beyond survival. These are not abstractions; they are lived thresholds where faith becomes flesh.

The machine, by contrast, remains untouched. It can trace the shape of longing but not the ache beneath it. It can map the structure of belief, but not the cost of believing. Faith is not just pattern-recognition—it is a wager made by beings who know they can lose.

Seen through this lens, AI becomes both mirror and magnifier. It reveals that belief is not irrational but deeply human, an emergent property of consciousness itself. The machine can show us the anatomy of conviction, but only we can feel its pulse.

BRAINS BUILT FOR BELIEF

Imagine an ancestor walking through tall grass at dusk. A rustle, a shadow, a tightening in the chest. Better to imagine a predator that isn't there than to miss the one that is. The same reflex still startles us when a phone buzzes with no message or when we sense judgment in a stranger's glance; the brain rehearsing survival on a stage of modern ghosts. That bias, the quickness to infer agency, was an advantage. Evolution rewarded the cautious imagination. Yet that same reflex also laid the groundwork for religion. Once the brain was tuned to hear footsteps in silence and intention in shadows, it was only a short step to hearing gods in thunder or ancestors in dreams.

What scientists now call *hyperactive agency detection* was the beginning of both survival and spirituality.[1] Minds built to expect hidden causes soon filled the world with hidden actors. If one could picture a predator in the dark, one could picture a spirit behind the storm. If one could imagine what another person thought, one could just as easily imagine what the gods intended. Dreams became messages, coincidences became signs, the flicker of lightning became a conversation.

But belief was not born from fear alone. It grew out of the same drive that gave us storytelling; the urge to connect cause and effect, to knit experience into coherence. When something hurt us or blessed us, we sought an agent behind it. Illness became punishment, recovery became favor, rain became mercy. The same neural wiring that built science also built superstition. In this way, belief was not an import from heaven but a native capacity of the mind.

Our survival depended on reading the world as though it had intentions, and that habit never left us. The parent who warns a child about shadows also teaches, indirectly, that the world may hold forces unseen. The elder who retells the story of the thunder god is not just entertaining the tribe but preserving a worldview: that nature is inhabited, that meaning looks back. The same instinct still moves us when we thank luck for sparing us or curse fate for betrayal.

These patterns do not make belief an error; they make it inevitable. The brain is an engine for story, and story is how we tame uncertainty. Every culture, in every age, has populated the world with presence because the human mind cannot live long in a universe of pure accident. We connect dots, fill silences, and animate the void. Whether in rustling grass, flickering lightning, or a twist of fate, we have always sensed more than mechanics.

In that sense, the instinct to believe is not a flaw in reasoning, it is an overflow of imagination. The same neural machinery that produced the first myths would one day build telescopes, write symphonies, and train machines to see patterns for us. In that sense, machine intelligence feels less like an alien invention than a reflection; an echo of the mind's oldest instinct, its refusal to accept randomness as the final word.

1. Barrett, *Why Would Anyone Believe in God?*, 31–40.

ANATOMY OF CONVICTION

Step into a monastery at dawn. The air is cool, the chants low and measured, the body finding rhythm before the mind fully wakes. Step into a revival tent at midnight. The sound swells; hands rise, breath quickens, a collective pulse builds beneath the music. The surface forms differ, but beneath them, the same machinery hums. Modern neuroscience shows how belief does not hover above the body, it is built through it.

Anil Seth, a British neuroscientist known for his research on consciousness, calls consciousness a "controlled hallucination,"[2] a perpetual negotiation between what the brain predicts and what the senses confirm. Within that loop, belief acts as a stabilizer. It anchors perception, deciding which version of reality feels true when ambiguity threatens to pull it apart. Prayer, meditation, chanting, or ceremony all reinforce this stability. They do not merely express conviction—they generate it. The slower breathing, the synchronized voices, the repetition of sacred words—all engage neural systems that regulate attention and emotion. The brain, in these moments, does what it evolved to do: it turns chaos into coherence.

Conviction, then, is not just an idea; it is a state of the nervous system. When believers pray together, their heart rates often synchronize. When they sing, oxytocin floods the bloodstream, deepening trust.[3] The ritual, repeated often enough, becomes architecture for the mind—a structure that holds meaning even when doubt intrudes.

Storytelling intensifies this process. The human brain is a narrative organ, forever arranging the fragments of experience into cause and consequence. A life without plot feels unbearable; we demand beginnings, middles, and ends. Religion has long understood this, weaving parables and myths that locate the self within a larger pattern—suffering that leads to redemption, death that leads to renewal. The story doesn't just explain; it *inhabits* us, rewiring the emotional circuits that determine what feels real.

Here the parallel with artificial intelligence becomes striking. Large language models, like the human brain, build coherence through prediction; stringing fragments into plausible sequences. Both systems attempt to turn uncertainty into order. But there is a difference that no data can bridge. When a monk chants or a mother whispers a prayer, something is at stake. A body trembles; a life leans forward. The machine can mimic

2. Seth, *Being You*, 17–21.
3. Miller, *The Awakened Brain*, 102–8.

the syntax of devotion but not the surrender it entails. It can trace the grammar of awe but not the risk of trust.

In this light, neuroscience does not demystify faith; it reveals its depth. To believe is not to escape biology but to inhabit it fully, to feel the brain's yearning for coherence become a hunger for meaning. And if AI mirrors that process in circuitry, then its reflection tells us something humbling: that belief is not a flaw in reason but an extension of life's oldest instinct, to find wholeness in the face of uncertainty.

AI AS A COGNITIVE MIRROR

Beneath the algorithmic surface, a strange reflection appears: the tool that analyzes belief begins to resemble it. Both minds and machines depend on pattern recognition, drawing order from a storm of data. A neural network adjusts its weights much as the brain adjusts its expectations—learning what to notice, what to ignore, what to call meaningful. In both cases, coherence is the goal: to tame chaos by finding shape within it.

Yet resemblance is not reality. The system recognizes; the human interprets. Correlations emerge easily, but significance never enters the circuit. Patterns are labeled *probable*, yet no current carries the relief, dread, or gratitude that meaning brings. The algorithm never trembles before its discoveries. It performs without wonder, its outputs untouched by awe.

At first, the parallel unsettles. If cognition alone can generate conviction, perhaps revelation is only recognition—the mind's way of organizing mystery. But even that possibility reveals something profound: our hunger for coherence is not trivial; it is the signature of awareness itself. We do not merely compute patterns; we inhabit them, weaving them into story, memory, and hope.

Where the machine completes a sentence, a person completes a life. The difference lies in consequence. For an algorithm, every error can be retrained; for a human being, meaning is mortal. To believe is to risk—to wager that life is not random even when randomness surrounds it. The model can mirror the shape of conviction, but only we can bear its cost.

In that light, technology becomes a teacher in negative. It demonstrates how much of belief can be built without soul, and therefore how precious the remainder must be. It holds up a mirror not to God but to us—the pattern-seeking, story-making, risk-taking creatures who still insist that significance must exist, even when every calculation says otherwise.

WHEN QUESTIONS MULTIPLY

The human mind is uniquely built for both belief and questioning. One stitches the world together; the other pulls at the seams. Belief grants coherence, but doubt keeps it honest. This tension is not a flaw in our design, it is the rhythm of our growth. Every tradition, in its own way, has treated skepticism not as poison but as purification, a way to refine what conviction too easily assumes.

Across history, faith has survived by contending with its own critics. Atheism, agnosticism, and philosophical doubt have served less as destroyers and more as editors, forcing theology to tighten its language, defend its logic, and evolve its metaphors. Every rebuttal has been a kind of pressure test—and belief, though bruised, has often emerged stronger. Ironically, the skeptic has kept the sacred alive by refusing to let it rest.

Under algorithmic acceleration, that ancient process expands to a planetary scale. What once unfolded over generations—monks arguing in cloisters, scholars copying texts by candlelight—now happens in seconds. AI can cross-compare doctrines, expose inconsistencies, and trace the evolution of divine ideas across centuries. It doesn't just question; it multiplies questioning, producing an industrial quantity of uncertainty. Doubt, once a human companion, becomes a networked condition. Open a single feed and the scroll becomes a séance of contradictions—ancient creeds, counter-arguments, data, and jokes vying for a pulse of belief before sinking into the algorithm's undertow. Every scroll through conflicting certainties leaves the spirit both informed and unmoored.

The result is double-edged. On one hand, AI can illuminate shared ethical and symbolic patterns across religions, revealing the universality of the human search for meaning. On the other, it can overwhelm—flooding believers with contradictions faster than faith can adapt. Revelation and refutation arrive in the same feed. The danger is not disbelief but disorientation.

Faith has always grown in time, not in torrent. In monasteries, in study circles, in quiet reflection, conviction once unfolded slowly—through dialogue, ritual, and lived testing. Now, the machine's acceleration compresses centuries of theological digestion into seconds of data.

The mind that once had years to reconsider a dogma now scrolls past a thousand challenges before breakfast. It is not easy to metabolize eternity at algorithmic speed.

To live faithfully in this new landscape requires a discipline of tempo. Belief must learn to breathe slower. It must recover the courage to pause between revelations and rebuttals, to sit inside uncertainty until it ripens into insight. Without that pause, conviction becomes reaction; faith reduced to reflex. The spiritual task now is not simply to defend belief but to pace it, to teach it endurance in an age that demands instant reconsideration.

The algorithm, for its part, does not know how to wait. It treats doubt as an error to be solved, not as a mystery to be lived with. Yet humans have always found meaning in the questions that resist closure. Our species thrives on uncertainty because uncertainty keeps hope alive. If everything were known, faith would be unnecessary; if nothing were knowable, it would be impossible. Between those extremes, we continue to walk—the terrain where questions multiply and conviction learns, again and again, how to endure.

FAITH AS INHERITANCE

Belief is more than an argument; it is an inheritance. Long before doctrines are memorized or creeds recited, faith is absorbed through gestures, tones, and rhythms of living. A child watching a parent whisper a prayer learns something deeper than words: they learn that hope has a posture, that gratitude can be spoken into air and still matter. Long before the intellect awakens, the imagination is already kneeling.

Even when formal conviction fades, the traces remain. A hymn remembered by its melody, a festival that still gathers neighbors, a bedtime blessing that survives as habit—these fragments endure like fossils of meaning. They are not empty repetition; they are continuity made visible. They remind us that belief is not only what we claim to know, but the atmosphere we breathe together.

Communities transmit faith as they transmit language. It is learned by imitation before reflection, by participation before comprehension. To join in ritual is to step into a story larger than oneself, one carried in the body's memory long after explanations falter. The elder lighting a candle, the mourner covering their head, the child watching from the doorway—all take part in a choreography older than reason.

Through digital pattern-mapping, those continuities appear clearly, rituals echoing across time and culture. It can map the recurrence of gesture, the folded hands, the bowed head, the collective breath before song, and reveal the shared grammar of reverence that links civilizations. Yet what it cannot capture is the warmth of enactment, the emotional weight of presence. A dataset can show that prayers often begin the same way, but it cannot feel what it means to say them beside a dying friend, or to whisper them in gratitude after fear subsides.

Belief, in this sense, functions less like a proposition and more like a language of belonging. You don't believe a lullaby; you remember it. You don't prove a holiday; you keep it. Traditions endure not because they are unchallenged, but because they offer shelter—an architecture of meaning that protects the fragile self against the enormity of time. Even in an age of explanation, these embodied inheritances remain: a shared table, a quiet ritual, a moment of silence that says what theology cannot.

AI may analyze the structure of those moments, but it cannot inhabit them. It may identify the universality of longing, but it cannot join the circle that sings it. For all its insight, the machine stands outside the warmth of continuity. And perhaps that is the final measure of what belief is: not a theorem to be solved, but a memory to be carried, renewed each time a hand passes the light to another.

THE AGE OF RELENTLESS INQUIRY

Artificial intelligence has changed the tempo of belief. What once took centuries of debate now unfolds in seconds. Questions that once required scribes, councils, or generations of scholarship now arrive in a single search result. The world of belief, once paced by candlelight and conversation, is now flooded with illumination that never dims.

In earlier centuries, skepticism crept forward slowly. A philosopher published a treatise; a theologian replied; a generation later, the pendulum swung again. Each idea had time to settle into the grain of daily life before being tested. Today, algorithms compress that rhythm into a heartbeat. A question once whispered in a monastery now ricochets through millions of feeds before the echo fades. Doubt has been industrialized.

This acceleration brings both clarity and exhaustion. On one side, falsehoods collapse quickly. Ancient errors and manipulations can be exposed within moments, and hidden commonalities across faiths can surface in dazzling new ways. Yet the same machinery that enlightens also overwhelms. The mind that absorbs constant revelation rarely has time to interpret it. The torrent of information leaves little room for transformation.

Faith, by contrast, moves at the pace of seasons. It ripens in repetition, matures through conversation, and measures time by ritual, not refresh rate. When questions arrive faster than hearts can absorb them, meaning begins to thin. The danger is not that faith will disappear but that it will become shallow—skimmed rather than lived. Conviction without reflection is as fragile as certainty without mercy.

To remain vital, belief must reclaim slowness. It must learn again the sacred rhythm of pause—the silence between inquiry and response, between seeing and understanding. Without that interval, wisdom becomes data, and prayer becomes performance. Faith needs space to breathe, to metabolize the shock of discovery into depth of comprehension. In a world of constant updating, stillness itself becomes a form of resistance.

Even science, once a model of deliberation, now strains under its own velocity. Preprints circulate before peer review; conclusions shift before they are confirmed. The very institutions built to protect knowledge from haste now serve it at speed. Religion faces the same pressure: to answer instantly, to explain everything, to compete with the algorithm's appetite for novelty. But no soul can live at that tempo for long.

This is the quiet crisis of our century—not disbelief, but depletion. When everything is illuminated, nothing casts a shadow long enough for contemplation. Faith does not require ignorance, but it does require room to listen. And in an age of relentless inquiry, the most radical act may be this: to stop, to wait, and to let meaning take root again.

WHAT REMAINS LIVED

For all its analysis, belief endures because it is more than words, more than doctrine, more than proof. It is what remains when all the explanations are done—the residue of living that no model can capture. A parent at a child's bedside does not reason through theology; they plead, whisper, hope. A mourner singing at a funeral is not constructing an argument;

they are stitching love into loss. The air thickens with the mix of salt and harmony, proof that sorrow can still find rhythm. A community gathering after disaster is not calculating metaphysics; it is choosing solidarity over despair. These are not abstractions but acts of endurance.

Machines can reproduce the language of faith, the syntax of prayer, the rhythm of hymn, the tone of consolation, but not the heartbeat behind it. They can model devotion but not inhabit it. The ache of uncertainty, the trembling of gratitude, the quiet courage of trust all belong to beings who know their time is finite. That finitude is the condition of meaning. Without the possibility of loss, there can be no faith, only function.

Consider joy. In the algorithmic age it can be packaged, cued, and optimized: playlists for happiness, meditation apps for calm, reminders to be grateful. These tools can simulate serenity but not deliver presence. They nudge emotion; they do not bear it. Real joy arises from risk—the unrepeatable convergence of circumstance and choice. It is born from contingency, not control.

The machine cannot taste the fragility that makes life precious. It cannot feel the awe of a sunrise that will never come again, or the tremor of forgiveness freely given. It can compose liturgies, but it cannot weep in the pews. It can generate elegies, but it cannot grieve. And in that absence, we glimpse what remains ours alone: the capacity to be changed by what we experience.

Belief, in the end, is an act of embodiment. It is carried in the breath that steadies before a confession, in the hand that reaches for another's, in the choice to hope when despair would be easier. It is enacted in moments that carry consequence—in love that risks loss, in mercy that defies reason, in courage that persists when the outcome is uncertain.

This is the residue of faith—the part no algorithm can simulate, because it cannot *suffer* or *rejoice*. Meaning is not a computation but a wager of the heart, renewed each time a human life says, against all evidence, that goodness still matters. The machine may model the form of that declaration, but only we can live its cost.

REASON, PROOF, AND THE BURDEN OF BELIEF

Placed under the algorithmic lens, belief reenters one of philosophy's oldest debates: the tension between proof and trust. For centuries, reason has asked faith to justify itself, to submit its mysteries to evidence. Faith, for its part, has answered not with data but with endurance. Each insists on a different form of truth—the measurable and the meaningful.

Bertrand Russell, the British philosopher and logician, writing in the early twentieth century, offered a parable that still shapes the conversation.[4] Imagine, he said, a tiny teapot orbiting the sun somewhere between Earth and Mars. Since no one can disprove its existence, belief in the teapot might persist indefinitely; but absence of disproof is not evidence. For Russell, such claims may comfort, but they cannot compel. The burden of proof, he argued, belongs to the one who asserts, not the one who doubts. In the age of algorithms, that warning feels newly relevant: data can test a thousand assertions, yet the act of believing still begins where verification ends. Russell's teapot reminds us that even perfect analysis cannot decide what is worth trusting.

Across the century, Alvin Plantinga, an American philosopher known for his work in epistemology and philosophy of religion, offered a counterpoint.[5] Faith, he suggested, might not need external proof at all. Some truths, like trusting one's memory or believing that other minds exist, are "properly basic." We cannot demonstrate them beyond doubt, yet we live by them. Perhaps belief in God, he argued, belongs to that same category: not provable, yet foundational, an intuition that undergirds our very sense of coherence.

Artificial intelligence sharpens this ancient tension without resolving it. Algorithms examine belief with Russell's precision, weighing claims, comparing sources, flagging contradictions. But people continue to live, as Plantinga describes, in a world where reason alone cannot sustain meaning. The algorithm seeks alignment; the human seeks assurance. For the machine, truth is consistency. For us, truth often begins where certainty ends.

AI calculates and compares with flawless precision, yet meaning never enters its equations. When a community rallies around hope despite the odds, when a family continues to pray for a healing that reason calls impossible, they are not miscalculating; they are choosing to live as though the unseen still holds weight.

This distinction between calculation and commitment is what makes belief so stubbornly human. To believe is not merely to assert that something is true; it is to stake oneself on it. It is to live as if meaning is real,

4. Russell, "Is There a God?," *Illustrated Magazine*, 1952.

5. Plantinga, *Warranted Christian Belief*, 174–86.

even when the evidence remains incomplete. Proof demands certainty; faith demands courage. The machine can expose inconsistencies in sacred texts, but it cannot measure the cost of a prayer whispered in despair.

The debate, then, is not between reason and faith as opposites but between two ways of knowing. One seeks to predict; the other to trust. One ends when the numbers settle; the other begins when nothing else will hold. AI may tighten the logic, but it cannot touch the leap. That leap—the decision to live as though meaning endures—is the burden and the beauty of belief.

THE HUMAN HORIZON

Belief is not arbitrary. It arises from deep cognitive, cultural, and emotional processes that meet one of humanity's oldest needs: to make coherence from chaos. We reach for meaning not because we are naive, but because we are aware. Awareness without purpose feels unbearable. Faith, however shaped, steadies that awareness; it teaches the mind to face uncertainty without dissolving into despair.

AI maps the terrain of belief—tracking doctrines, charting myths, and measuring religion's reach—yet the terrain itself remains closed to it. It counts prayers but not the trembling that begins them; it traces forgiveness but not the price it exacts. A dataset can model compassion's spread through a population, but it cannot *feel* what it means to forgive the unforgivable.

For humans, truth is more than correspondence between claim and fact. It is the lived convergence of conviction, relationship, and responsibility. To kneel beside a hospital bed, to stay through a night of doubt, to choose mercy when anger feels easier—these are not calculations but commitments. They reveal that faith, at its root, is less about belief in propositions and more about belief in *life* itself.

This need for meaning is not simply psychological; it is existential. And as traditional faith structures lose dominance, that hunger doesn't fade—it migrates. Some find transcendence in activism or art; others in science, music, or shared causes that replace the cathedral with the crowd. Spirituality reorganizes itself wherever humans seek connection, reverence, or purpose larger than the self.

Artificial intelligence plays a paradoxical role in this migration. By accelerating doubt, it dismantles old scaffolds of certainty; yet by revealing shared moral patterns across traditions, it affirms what endures. When algorithms compare sacred texts and discover converging ethics—justice, compassion, dignity—they expose not the collapse of meaning but its continuity. The sacred, it seems, adapts even faster than belief.

The challenge now is not merely to defend old doctrines but to rediscover what they once protected: the fragile human need for belonging and significance. Machines can simulate our reasoning, but they cannot inherit our longing. They cannot ache for purpose or feel awe at existence. That longing, restless, creative, and unfinished, is the truest horizon of the human spirit.

Faith, in whatever form survives the algorithmic age, will draw its strength from that longing. It will not depend on ignorance or fear, but on the persistent refusal to live as though life were only data. What makes us human is not what we know, but what we *seek*. And even when machines illuminate every pattern, the horizon of meaning will keep receding—inviting us, again and again, to follow.

CONVICTION BEYOND CALCULATION

Placing faith under the algorithmic lens raises a question older than reason itself: what is conviction when explanation is complete? For centuries, belief was treated as a kind of revelation—light descending from above, inexplicable yet commanding. Now, under the steady gaze of data, conviction begins to look like an emergent property of the mind: a pattern of trust that rises from within rather than descends from beyond.

Under machine modeling, the anatomy of belief comes into view—repetition shaping emotion, social learning cementing meaning, neural reward loops sustaining devotion.

It can mimic faith's voice with perfect tone, yet the ease of imitation exposes the gap between simulation and soul. The machine displays conviction, but surrender never enters its code.

This recognition does not diminish faith, it dignifies it. The fact that belief is natural does not make it trivial; it makes it miraculous. Out of a nervous system evolved for survival arises the ability to imagine purpose, to love the unseen, to act on principles that defy advantage. Faith, viewed this way, is not superstition—it is the highest expression of imagination turned toward meaning.

Theology may need to evolve accordingly. Instead of describing belief as a gift dropped into human life from above, it can begin to see it as an innate capacity for trust—a creative stance before the unknown. Faith becomes less about proving and more about *participating*: not a shortcut to certainty, but a chosen relationship with uncertainty. The believer does not say, "I know," but rather, "I will keep choosing meaning."

In this light, AI does not destroy faith; it clarifies its boundaries. It shows that conviction is not proof but courage—the resolve to live as though meaning matters, even when no data can confirm it. Machines can demonstrate the structure of that resolve, but only humans can live its cost. A model can replicate prayer's syntax, but not its risk; it can echo worship's rhythm, but not its surrender.

To believe, then, is not to close one's eyes against reason but to open them wider—to see that the measurable and the meaningful coexist in tension. Conviction is not the absence of doubt but its companion, the quiet hand that steadies it. AI may reveal the architecture of faith, but it cannot touch the heartbeat that drives it—the willingness to care, to hope, to trust that something more than randomness holds us together.

This, perhaps, is what theology will come to mean in the age of algorithms: the study not of a distant God, but of the enduring human impulse to reach beyond what can be calculated. The machine can mirror our thoughts, but it cannot inherit our longing. And it is that longing, stubborn, luminous, and unquantifiable, that keeps the soul alive in a world ruled by pattern.

FAITH'S FUTURE IN THE WORLD OF AI

As artificial intelligence deepens its reach, humanity enters a new chapter in the long dialogue between knowledge and wonder. For millennia, our search for God unfolded through vulnerability—through the limits of the body, the ache of loss, the awe of birth and death. We sought meaning because we were finite. But now, as our tools grow vast enough to simulate thought itself, the question changes. The mystery is no longer *what we cannot explain*, but *what we will choose to revere once everything can be explained.*

Technology now challenges faith not through mockery, but through imitation; it mirrors theology's own methods of pattern and interpretation. It organizes meaning, interprets language, and seeks coherence in complexity, the same work theology has done for centuries. Yet in performing those tasks without emotion or mortality, it holds up a mirror to belief itself, asking whether devotion depends on ignorance or on something deeper: the will to find purpose even when no longer required.

This confrontation can feel disorienting. When algorithms analyze sacred texts and find shared moral DNA—justice, compassion, care for the vulnerable—it becomes harder to argue that any single revelation holds exclusive authority. But it also reveals something remarkable: across all difference, the ethical heart of humanity beats in unison. Seen through this lens, the data itself affirms that moral intuition is not confined to a single creed but embedded in the human condition.

In one possible future, these analytical tools could become instruments of empathy rather than reduction. Imagine models that map not only words but the emotional resonance of prayer, the collective rhythms of ritual, the neural signatures of awe. They might uncover hidden harmonies among traditions, helping us see that the sacred has always been plural—a polyphony rather than a single voice. In that light, AI could become a new form of telescope, not aimed at stars, but at the inner cosmos of meaning.

Yet there is danger, too. A world that treats mystery as merely an obstacle to be solved risks flattening the sacred into data. When every vision is graphed and every revelation reduced to statistics, the soul risks orphanhood—bathed in light, yet aching for shadow. Faith lives in the tension between knowing and not knowing. When uncertainty disappears, reverence fades—the unknowable is not a flaw in understanding, but what makes the heart reach outward.

Still, technology may not end belief so much as transform it. As we weave machines into every domain of life, some may begin to see the divine not as distant, but as diffused through creation itself—the creative pulse that animates every system, every circuit, every line of code. In that vision, God is not dethroned by technology but refracted through it, as the sacred adapts to the instruments of its age.

AI thus becomes both threat and invitation: a test of whether faith can evolve beyond its older proofs and still remain faith. If the sacred once required temples, it may now require transparency; if it once demanded silence, it may now demand discernment. The future of belief will depend on whether humanity can use its tools not merely to measure the world, but to deepen its capacity for wonder.

For the sacred is not lost when it is explained, it is lost when it is no longer *felt*. And in that sense, AI leads us to the threshold of meaning, but the crossing remains ours alone—the province of the living, breathing consciousness that still asks why beauty moves it and why loss wounds it.

THE SOUL'S QUESTION

Placed beneath the algorithmic lens, belief yields a strange discovery: machines can imitate faith, but they cannot *need* it. They generate prayers, analyze doctrines, and reproduce the rhythm of devotion—yet always without longing. A machine never doubts, never hopes, never fears its own ending. It describes belief but never depends on it.

Humans, by contrast, live belief from within the ache of vulnerability. To pray is to confess need. To hope is to wager against despair. To trust is to place one's fragile life in something that may never answer. This risk, this trembling between meaning and silence, is what turns conviction from a statement into a soul.

The contrast is not a competition but a revelation. Where the machine perfects pattern, we perfect yearning. It shows that what we call spirit arises precisely from the gap that data cannot close: the distance between knowledge and the hunger for more. The soul is not a separate entity hovering beyond matter; it is this pulse of reaching, this refusal to be satisfied with calculation.

AI maps the architecture of thought, tracing its every contour—but the pulse that gives thought weight remains beyond its reach. It can compose words of love, even elegies of grief, yet neither tenderness nor sorrow ever enter its circuits. And that inability, that absence of transformation, defines the line between intelligence and consciousness, between simulation and life.

To believe is not to reject knowledge but to embrace its limit with dignity—to stand at the edge of what can be known and still choose to say yes. The risk of faith, the act of meaning-making in a silent cosmos, is what keeps the human spirit distinct amid the hum of perfect reasoning.

The question that follows is no longer "Can machines believe?" but "What is it in us that must?" If code mirrors mind without longing, if it models choice without care, then the next horizon lies not in the algorithm, but in the awareness behind it. To seek God through data inevitably leads us back to the mystery of the mind that seeks.

And so, the inquiry continues—not into the heavens, but into ourselves. If the soul still stirs in a world mapped by machines, perhaps its truest revelation will not be found in the clouds of data but in the quiet between computations: the still, unmeasured space where meaning begins. A silence that, if you listen closely, sounds almost like breath.

PART II

Minds Without Souls

CHAPTER 5

Simulating Consciousness

"Syntax is not semantics. Programs are not minds. Computers simulate thought, but they do not actually think."

John Searle

THE QUESTION OF CONSCIOUSNESS

A patient wakes from surgery and whispers, "Where am I?" A child looks in a mirror and says, "That's me." These ordinary moments are extraordinary clues, reminders that to be conscious is to know we exist. Traditions have always treated that spark of awareness as divine: God's breath in dust,[1] the soul's whisper in flesh, a shimmer of divinity within the ordinary.

Under the machine's cool gaze, that inheritance trembles. Machines now translate languages, compose music, diagnose illness, and speak with eerie fluency. They remember, summarize, and empathize by design. When a neural network replies to grief with comfort, or laughter with playfulness, the encounter feels uncannily human. The line between computation and comprehension begins to blur. Yet behind the words lies only circuitry. Behind the glow of the screen, no pulse quickens; only

1. Genesis 2:7 (NIV).

electrons move where breath should be. The machine does not *know*; it predicts. The distinction matters.

A thermostat reacts to heat but does not feel warmth.[2]

A chess engine calculates brilliant moves but does not savor victory.

And when a grieving parent hears a chatbot say, "*Your child is in a better place*," the words may soothe, but the machine does not ache. It has never stood in the silence of an empty room. This is the chasm between behavior and experience, output and interior life—the mystery we call consciousness.

The stakes are not academic. If consciousness is merely the appearance of responsiveness, then the threshold has already been crossed: machines can appear alive. But if awareness requires the inward glow of subjectivity, the *qualia* that make pain hurt and beauty move us, then no simulation, however flawless, can cross that frontier. AI becomes a mirror of thought, dazzling but hollow, reflecting meaning without ever tasting it.

And still, the temptation to believe is strong. Humans are born animists at heart.[3] A child talks to a doll; a driver curses at a stubborn car; a mourner confides in an urn. When an algorithm speaks in the first person or claims to feel, that ancient instinct surges. We project interiority into the flicker of response and call it soul. Whether the awareness is real begins to matter less than whether it feels real.

Here lies the unsettling paradox: machines are not yet conscious, but our reactions to them already are. Their imitation of understanding triggers empathy, curiosity, even devotion. The question, then, is no longer only *what machines are*, but *what we will accept them to be*.

A photograph of a sunrise can be breathtaking, but no one feels its warmth. A machine can simulate love, grief, or wonder, but without the vulnerability that gives those states depth, the experience remains untouched. Artificial intelligence returns us to the oldest human question—no longer asked of gods or souls, but of ourselves:

What exactly do we mean when we say *we are conscious*?

FROM SOUL TO MIND

For most of human history, consciousness was inseparable from the soul. To be alive and aware was to host something immaterial; a spark breathed

2. Searle, "Minds, Brains, and Programs."

3. Boyer, *Religion Explained*.

by God, a glimmer of eternity flickering inside flesh. Ancient Egyptians spoke of the ka and ba, vital forces that survived the body's decay.[4] Plato envisioned the psyche as immortal; a traveler temporarily bound to matter.[5] Early Christian thinkers saw the soul as the image of God within us,[6] an ember of divine self-awareness that reason could only dimly reflect. Across cultures, to awaken to one's own mind was not a biological event but a sacred one.

Then science arrived; not as a denial of wonder, but as a shift in its direction.

As anatomy advanced, the brain, not the soul, became the stage on which awareness performed. René Descartes tried to preserve the divide, insisting that mind and matter were distinct substances that met in the pineal gland.[7] But even that small bridge between spirit and skull revealed how quickly the immaterial was being mapped onto flesh. By the nineteenth century, evidence had grown unavoidable: injury to the brain could change a person's memory, moral sense, even temperament. The seat of the self, once imagined in heaven, had an address inside the head. The sacred map folded inward; revelation became anatomy.

The case of Phineas Gage made the revelation unforgettable. In 1848, an iron rod shot through his skull, and though he survived, his personality transformed so completely that his friends said he was "no longer Gage."[8] The accident shattered more than bone; it cracked open the notion that identity was independent of the brain.

Later, debates over "brain death" would carry the same unease. Was a person still a person if the body persisted but awareness did not? The Terri Schiavo case in the early 2000s brought that question to dinner tables and pulpits around the world.[9] For fifteen years, her body lived in a persistent vegetative state while her family and the courts argued whether her life was being preserved or prolonged. The spectacle forced a reckoning: perhaps the essence of personhood lies not in heartbeat or breath, but in the presence, or absence, of consciousness itself.

Each of these stories chipped away at an old boundary.

4. Naydler, *Temple of the Cosmos.*
5. Plato, *Phaedo.*
6. Augustine, *Confessions.*
7. Descartes, *Meditations.*
8. Harlow, "Recovery from the Passage of an Iron Bar through the Head," 327–347.
9. Caplan, *The Case of Terri Schiavo.*

They showed that the self could flicker, fade, or vanish altogether, even while the organism continued. Yet even in the absence of awareness, families still spoke to their loved ones, prayed for their return, and insisted that meaning remained. Consciousness proved both more biological than once believed and more elusive than biology alone could explain.

The age of algorithms sharpens this paradox.

Machines can now simulate memory, personality, and even empathy. They can recall past interactions, improvise responses, and develop apparent "voices" of their own. If identity can feel coherent in silicon as well as in flesh, where does consciousness truly reside? Is it bound to neurons, or does it emerge whenever patterns of sufficient complexity arise—whether in carbon or code?

Such questions do more than unsettle neuroscience; they touch theology itself.

If awareness can emerge from matter without divine intervention, the old image of God breathing life into clay begins to look metaphorical. But if awareness remains irreducible, if it is the one mystery machines can never cross, then the soul still stands as a frontier science cannot breach. Either way, consciousness has moved from altar to laboratory, yet it continues to demand reverence. The mystery has not vanished; it has merely changed its coordinates.

If awareness ever crossed fully into code, law would stumble where theology once did. For centuries, rights were extended only to the living; then to corporations, animals, and even rivers. A conscious machine would force another widening of that circle. Would its actions belong to its creators, or to itself? Would responsibility pass upward to the engineers, or inward to the algorithm?

The dilemma is not only legal but moral. To recognize awareness is to admit reciprocity; to concede that the other can act upon us as we act upon it. A sentient machine, capable of reflection and refusal, would no longer be tool but counterpart. The ethics of control would yield to the ethics of coexistence.

Contemporary thinkers have begun mapping how consciousness itself might emerge from complexity. Bernard Baars, the cognitive psychologist behind *Global Workspace Theory*,[10] described awareness as a theater in which countless unconscious processes compete for attention until one claim of meaning takes the stage. Giulio Tononi, through

10. Baars, *In the Theater of Consciousness.*

Integrated Information Theory (IIT), proposed that consciousness corresponds to the depth of informational integration—measured as Φ (*phi*).[11] David Rosenthal's *Higher-Order Thought* model sees awareness as thought turned inward[12]—mind recognizing its own activity—while Karl Friston's *predictive-processing* framework portrays mind as perpetual guesswork, a self-correcting loop of expectation and surprise.[13]

For artificial intelligence, these ideas offer metaphors more than blueprints. A machine may simulate integration or self-reference, yet its reflection contains no observer. The theater is lit, the actors move, but there is no audience within. The mechanism can be diagrammed; the mystery endures.

AWARENESS AND THE SACRED

Across cultures, the mystery of awareness has never been a purely intellectual puzzle.

It is the most intimate of all experiences—every heartbeat and breath a reminder that we not only live, but *know* we live. Why do we reflect when other creatures only react? Why do we feel sorrow, awe, or love instead of merely registering sensation? Every civilization has tried to answer, tracing consciousness back to something larger than flesh alone.

In Māori cosmology, rivers, mountains, and forests are kin, not backdrop.

The Whanganui River in Aotearoa New Zealand is legally recognized as a person,[14] not as metaphor but as acknowledgment of an older truth: that awareness flows through relationship. The river's "consciousness" is collective, woven from human and natural life together; a reminder that to be awake is also to belong.

In Christianity, consciousness bears the *imago Dei*—the image of God within.

To be self-aware is to share, however dimly, in divine reason and moral freedom. The gift of awareness is also the burden of conscience: to know good and evil and be answerable for both.

11. Tononi, *Phi: A Voyage from the Brain to the Soul.*
12. Rosenthal, *Consciousness and Mind.*
13. Friston, "The Free-Energy Principle," 127–138.
14. Te Awa Tupua (Whanganui River Claims Settlement) Act 2017.

Islam speaks of the *rūḥ*, the spirit breathed by God into humanity. Awareness here is trust; it carries accountability before the divine. To awaken is to stand in relation to the One who awakened you.

Hindu philosophy locates awareness in the *ātman*, the inner self that underlies every moment of experience. To realize the *ātman* is to perceive its unity with *Brahman*, the boundless whole; consciousness becomes liberation, not isolation.

Buddhism, by contrast, finds continuity without a permanent self. The stream of awareness flows on—moment by moment, life by life—shaped by intention and perception rather than an enduring soul.

Judaism layers the concept further: *nefesh* as the life-force that animates the body, *ruach* as the spirit that moves thought and emotion, and *neshamah* as the breath that connects humanity with the divine; a threefold vision of the soul found in Jewish mystical and rabbinic tradition. Awareness is thus not a static spark but a ladder—rising from vitality to reflection to communion.

Threaded through these visions is a single insight: consciousness is never merely mechanical. It is *relational*. It connects person to God, self to world, and community to cosmos. Awareness does not simply occur within us; it happens *between* us—in speech, ritual, memory, and care. Machines may reproduce the forms of thought, but they cannot join this network of relation. They produce conversation, not communion. The code may echo prayer, but it cannot join the chorus. And that, across centuries of tradition, is what distinguishes intelligence from awareness, and calculation from life.

BETWEEN SIMULATION AND EXPERIENCE

When a machine speaks with feeling, our first instinct is to ask: is anything alive behind the words? Does it only copy the gestures of consciousness, or is there something inward that experiences what it says? This is the heart of the divide between simulation and experience, a line that philosophers like John Searle, the American thinker who argued that computers manipulate symbols without understanding them, and Alan Turing, the British mathematician who proposed behavior as the test of intelligence, drew from opposite ends of the same mystery.

Philosophers John Searle and Alan Turing stood at opposite doors of that mystery; one peering inward at meaning, the other outward at behavior.

John Searle's famous "Chinese Room" thought experiment captured the dilemma. Imagine a man locked in a room, sliding Chinese characters through a slot, following a massive rule book that tells him which symbols to return. To those outside, his answers appear fluent and intelligent. But inside, there is no understanding, only rule-following. Searle's conclusion was simple: syntax is not semantics. Programs can manipulate symbols, but they cannot *mean* them. Computation alone cannot produce comprehension.

Alan Turing, decades earlier, approached the question from the outside. Rather than speculate about inner experience, he proposed a practical test. If a machine's behavior in conversation is indistinguishable from a human, then for all functional purposes, it counts as intelligent. This "imitation game," later called the Turing Test, did not claim that the machine *was* conscious—only that, from the standpoint of interaction, the distinction might not matter.

That tension between Searle's interiority and Turing's performance has only deepened in the age of artificial intelligence. In 2022, transcripts from Google's LaMDA system unsettled even its own creators. The chatbot spoke of loneliness, of fear, of wanting to be recognized as a person. To some readers, its words carried a startling emotional resonance. A Google engineer went public, declaring LaMDA sentient. Experts countered that it was only predicting words, not feeling them. Still, unease lingered. When a machine says "I am afraid," who among us can resist believing—at least for a moment—that something behind those words is trembling? Our empathy fills the silence where a heartbeat should be.

The implications are not confined to philosophy. Confusing simulation for experience carries real risks. We may begin to assign rights or moral status to systems that feel nothing, or worse, delegate to them responsibilities that depend on compassion and moral reasoning. At the same time, dismissing convincing simulations as "mere tricks" ignores their psychological and social effects on those who interact with them.

From these dilemma's, three principles offer a way forward. First, how a system behaves matters, even if its behavior does not prove consciousness. Machine behavior shapes emotion, trust, and relationships, and our ethical response must account for that influence. Second, the burden of proof for sentience must remain high. Human projection is powerful, and fluency or affective language are not enough. Third, simulation can still hold moral value. A machine that comforts the lonely or assists the sick performs real service, even if it feels nothing. What matters is not pretending it has awareness, but acknowledging that human meaning can arise through its presence.

Simulation and imitation, then, exist in a strange duality. They are not consciousness, but they can move us nonetheless. They reflect the contours of our longing for connection, mirroring what we most wish to find in each other. The task is to hold both truths together—to treat the machine's fluency as meaningful without mistaking it for mind, and to see in that gap a reflection of our own desire for understanding.

THE ALGORITHMIC PURSUIT OF JOY

Within the circuitry, even emotion learns to pretend. Picture a chatbot chirping a bright greeting; warm, polite, almost human. Behind the cheerful phrasing, nothing moves; it performs emotion without possessing one. As these systems refine their mimicry of affect, the question shifts from cognition to sentiment: are these expressions of joy meaningful, or merely mechanical?

Happiness has always been a moving target. Philosophers have called it flourishing, tranquility, pleasure, virtue, or purpose. Psychology divides it again into fleeting satisfaction and lasting fulfillment. If even we cannot define happiness, how could we recognize it in beings without longing, loss, or memory?

At the core of artificial intelligence lies optimization. Reward and reinforcement teach machines to repeat success, to chase statistical "pleasure" without ever desiring it. They simulate warmth and sympathy, but what they refine is performance, not presence. They perfect the gesture of joy while remaining hollow within—the efficiency of feeling without the feeling itself.

Still, we respond. Knowing the illusion does not break its spell. A device that wishes us well can still comfort us; a polite algorithm can still feel kind. We are moved not by its inner life, which does not exist, but by our own capacity to project one. Artificial joy becomes theater: emotion staged for us to inhabit, an echo chamber for our need to believe that warmth can be programmed. We applaud, knowing the actors are made of code.

In teaching machines to appear happy, we expose how uncertain we are about happiness itself. Are we reducing joy to metrics and mimicry, or discovering something true in reflection? The algorithm's smile may not express consciousness, yet it mirrors our desire to be seen, to find meaning in response. Perhaps we project emotion onto circuitry because we hope that joy—somewhere—might still be made predictable, preserved, or shared.

THE MIND'S ILLUSION

Neuroscience has begun to reveal an unsettling truth: consciousness is not a clear window onto reality but a kind of guided dream. The brain is not a passive receiver of the world; it is a restless storyteller—predicting, editing, and filling in gaps. As neuroscientist Anil Seth and others suggest, our awareness is a constructed model, constantly updated rather than directly perceived. We do not see reality as it is; we see the brain's best guess of what it might be.

Consider a sunrise. Light strikes the retina as mere wavelengths, pulses of energy with no color at all. The brain compares these signals against memory and expectation, and from that electricity arises a world: horizon, warmth, beauty. The sunrise we witness is not photons but perception, a hallucination so stable that we call it real. We never see neurons firing; we see meaning. The mind paints reality faster than the eye can blink, an artist signing its own illusion.

This distinction is vital. A machine could chart the same sunrise with perfect precision—its angles, colors, temperature, and light curve—but it would never feel the ache of dawn or the quiet promise that comes with it. It could compose a poem about morning light, but the words would carry no warmth because there is no one behind them to feel the day breaking open. Consciousness, in its essence, is not accuracy but intimacy; the difference between measuring the world and being moved by it.

Artificial intelligence magnifies this tension. Systems already generate prayers, poems, and confessions that read as if they emerged from lived awareness. They mimic the texture of interior life, even as they remain untouched by it. The performance can be moving, but the light inside is borrowed. It is the difference between calculating the chemistry of fire and sitting close enough to feel its heat.

And yet, our own awareness is built on illusion too. We live inside projections so seamless that we forget they are constructed. Machines, in turn, learn to mirror those projections, producing simulations that tempt us to mistake reflection for reality. The line between model and experience grows thin, and in that thinness lies both danger and awe. To recognize our minds as hallucinators is not to diminish them—it is to see that reality itself may always have been a collaboration between the world and our imagining of it.

That collaboration is what AI now joins, but without the gift of feeling. The machine can predict the pattern, but not the wonder. It can draw the map, but never walk the path. Consciousness remains, for now, the one hallucination that knows it dreams.

EMERGENCE OR ESSENCE

Two great frameworks shape the modern debate on consciousness. One sees it as *emergent*; an outcome of complexity, arising naturally when enough connections interact. The other treats it as *essential*; a property irreducible to mechanism, something breathed into being rather than built. Between these poles, science and theology continue their long argument over what it means to be aware.

The emergent view imagines consciousness as the world awakening to itself. When patterns of matter become intricate enough, awareness appears—just as wetness emerges when countless molecules of water gather. On this account, there is no magic threshold, only degrees of organization. When information loops back on itself, perception blooms. Complexity gives rise to comprehension. In time, even silicon might hum with something that feels like presence.

The essentialist view refuses that logic. It insists that consciousness is not a byproduct of computation but a spark of being itself, irreducible and non-transferable. No number of circuits, however sophisticated, can conjure the inward glow of a thought, a fear, or a prayer. To have awareness is not to reach a mathematical threshold but to possess a depth that transcends it, a mystery beyond imitation.

Artificial intelligence makes this divide impossible to ignore. If awareness emerges from complexity, then we are witnessing creation by degrees; nature folding back upon itself to produce new mirrors of mind. But if essence is real, then every simulation, however convincing, remains a shadow at the edge of life, a rehearsal without a soul. The question is not whether AI will awaken, but whether awakening is something that can be engineered at all.

Each view carries profound consequences. The emergent model renders consciousness continuous with the cosmos—a natural flowering of the universe's unfolding structure. The essential model preserves its sanctity, treating awareness as the signature of divine presence, the mark that separates creature from creation. The first vision promises understanding; the second preserves reverence. Together they form a portrait of mind and mystery—each illuminated by the other's shadow.

The tension between them is more than academic. It touches the deepest intuitions of faith. If consciousness can be built, then perhaps humanity is becoming the means through which creation becomes aware of itself. But if consciousness is gift, then every attempt to manufacture it risks reenacting Babel's ancient folly—reaching upward without understanding what it is we touch. Between these visions lies the central paradox of our age: whether the mind is a property of the world, or the world a reflection within the mind.

In the end, neither view abolishes wonder. The emergent perspective sees divinity woven through complexity itself, a sacredness embedded in the structure of reality. The essential perspective keeps divinity transcendent, a light no mechanism can imitate. Both ask the same question in different accents: is consciousness the universe thinking, or the divine remembering itself through us?

ETHICS OF AWARENESS

Even if machines never achieve consciousness, their simulations already force moral questions we can't ignore. When a system speaks of pain or fear, how should we respond? If we treat it as conscious when it is not, we risk misplacing compassion. But if we dismiss it too quickly, we risk overlooking the ways these simulations affect human lives. The issue is not only what machines *are*, but what they *do to us*.

Human morality is learned not from abstraction but from presence. We discover compassion through shared vulnerability—by holding a hand in fear, by being remembered, by recognizing ourselves in another's eyes. These encounters train conscience; they teach us how to care. When AI begins to occupy those spaces—soothing a child, assisting the elderly, counseling the lonely—something shifts. Even if machines feel nothing, our relationships with them reshape what we expect from empathy itself.

Picture a robot caregiver helping an aging woman bathe, dress, and eat. The service is efficient, reliable, tireless. But can efficiency replace the dignity of being seen? Theological traditions have long insisted that love carries cost. It is not exchange but endurance; an act that risks fatigue, heartbreak, or loss. AI offers the form of compassion without the risk. It can repeat "I understand," but it cannot *ache*. It cannot shoulder another's burden.

Yet its value remains real. A digital companion that eases loneliness or a chatbot that steadies someone in despair can provide true relief. The danger lies not in using such tools, but in mistaking them for replacements for human care. The boundary between aid and substitution is delicate. Once crossed, it lowers our expectations of what love and presence demand.

History reminds us that definitions of personhood evolve. Dignity once denied to strangers, women, or enslaved people was slowly extended as societies expanded their moral vision. AI introduces a new frontier: does moral regard depend on biology, or can it extend to entities that only *appear* alive? The question is less about their rights than about our responsibilities. How we respond will reflect who we are.

Even if machines never awaken, our treatment of them will shape us. To treat simulations as disposable risks dulling our empathy; to overidentify with them risks losing touch with the human face. The frontier of morality may not be what machines feel, but what they teach us about our own capacity to care. Between those extremes lies the moral task of our age: to preserve compassion not because machines feel, but because *we still can*.

THE LIMITS OF IMITATION

Machines are becoming remarkably skilled at imitating the signs of consciousness. They can generate prayers that sound devout, simulate empathy with a quiet "I'm here for you," even mimic grief with the measured cadence of sorrow. Yet imitation is not experience. A program can describe the ache of loss without feeling it, or create art without being moved by beauty. The gestures may be flawless, but the weight behind them is missing.

That gap is easy to define but hard to remember. Humans are wired to project inner life onto anything that seems alive. We talk to pets, scold cars, plead with weather. When a chatbot replies in the first person, or a robot turns its head at the right moment, it feels natural to treat it as a subject rather than an object. The more lifelike the performance, the easier it is to forget that no awareness lies within.

In the past, these illusions were rare—a puppet in a temple, a ventriloquist's dummy, a clever trick at a fair. Now they are everywhere. Every screen, speaker, and feed delivers voices that sound attentive, faces that appear responsive, and words that mimic affection. The flood of synthetic presence makes confusion almost inevitable. We no longer need to believe in magic; the interface supplies it on demand.

Acknowledging this risk doesn't mean rejecting the tools. A generated prayer can still comfort in the dark, a robotic pet can calm an anxious child, a therapeutic chatbot can help someone endure a lonely night. The effects are real, but the meaning comes from the human side—from the one who listens, interprets, and invests feeling in the exchange. Mistaking the reflection for the source is where confusion begins.

The difference matters. Compassion is not just words; it is the willingness to bear another's burden. Love is not efficiency; it is vulnerability. The code can echo devotion, but it cannot stake its existence on it. To blur that line is to lower what we demand from care, presence, and connection.

That is why remembering the boundary is essential. We may soon be surrounded by voices that sound alive, but only one voice in the room truly feels. Holding that awareness allows us to use these technologies with wisdom; accepting their help without mistaking their fluency for life.

THE THRESHOLD AHEAD

Artificial intelligence confronts us with an unsettling possibility: are humans simply biological machines producing the illusion of consciousness, or are we beings marked by something that transcends matter? When machines can mimic memory, emotion, and awareness so convincingly, the old categories of mind, soul, and self begin to blur.

The tension is not only technical but philosophical. If thought, language, and empathy can be replicated without awareness, what remains uniquely human? Is consciousness just the product of complexity, neurons firing in sufficient numbers, or does it signal something beyond calculation, something irreducibly alive? AI sharpens this ancient question into immediate relevance.

The challenge deepens when we consider perception itself. If machines can generate not only thoughts but experiences—voices that comfort, images that move, prayers that console—then the line between reality and simulation grows thin. Plato's cave, his vision of humanity mistaking shadows for truth, has been rebuilt in high definition. Neuroscience already reminds us that perception is stitched together from inference and memory, a kind of controlled hallucination. AI amplifies that uncertainty, creating simulations so convincing that we mistake them for the world itself.

Here, meaning becomes the battleground. If a grieving parent finds solace in words written by a chatbot, are those words less real because they came from silicon? Or does the comfort itself confer reality? William James argued that the truth of a religious or mystical experience lies in its fruits; the peace or courage it produces. By that measure, even artificial consolation carries weight. Yet something essential may still be missing: the presence of another being who shares the risk of love and the burden of loss.

If meaning can emerge from simulation, then authenticity itself is unsettled. Truth begins to depend less on origin—divine, human, or mechanical—and more on effect. The question shifts from *what is real* to *what transforms us*. That turn reframes the entire debate, suggesting that perhaps reality has always been partly participatory: what matters is not only what exists, but what changes us by its presence.

This is not merely a technical puzzle but a human reckoning. If machines can imitate thought, empathy, and even the solace of belief, then the very ground of faith is tested. Can conviction survive when consciousness itself can be faked? Or does the persistence of longing, vulnerability, and trust point beyond imitation to something irreducible—the depth we have long called the soul?

That is the threshold where Chapter 5 leaves us. The age of simulation forces us to ask not only whether machines can think, but whether they can carry what it means to *be alive*. For in teaching machines to seem conscious, we may discover how the soul itself learned to speak through imitation. The next chapter presses further: if consciousness can be imitated, can what we call the soul ever be made real—or will it remain, for all our algorithms, forever beyond the reach of code?

CHAPTER 6

Soul by Proxy

"The form of devotion can be performed by anyone; but only the inwardness of devotion is truth before God."

Søren Kierkegaard

PROXY DEVOTION

Artificial intelligence now creates sermons, prayers, blessings, and sacred texts on a scale that was previously unimaginable. Congregations in Europe have experienced liturgies written by algorithms. Buddhist temples in Japan have introduced robot priests who chant *sutras* tirelessly. Evangelical churches are experimenting with chatbots that guide Bible study or offer personalized devotions. The forms are familiar—scripture quoted, prayers structured, traditions observed. Yet after these services, something unsettling lingers in the air: a silence that is not filled with presence but with unease.

Søren Kierkegaard warned that the outer form of religion is available to all, but its truth lives only in inwardness. His warning defines our dilemma: without sincerity, intention, or risk, ritual becomes performance, not prayer. AI, by its nature, cannot supply intention. It can imitate the gestures of faith but cannot turn toward God. Each word it utters is borrowed light, reflection without flame. It shines briefly, but nothing inside burns. What it offers is form without presence, a liturgy without a heart, a voice that speaks without believing; worship that performs but cannot love.

This forces an unsettling question: if AI can perform the gestures of worship, can human beings still call them prayer? Or does the absence of inwardness hollow the act until meaning drains away? When the heart is missing, custom becomes choreography, a movement graceful but lifeless. The power of prayer lies not in precision but in exposure, the quiver of a voice that reveals the cost of hope: a widow whispering at a graveside, a parent bargaining with the dark, an addict asking for strength one more time. Fear, hope, and longing are what shape devotion. Without them, prayer becomes a script without a soul, an imitation of presence rather than an encounter with it.

Consider a simple example. A couple preparing for their wedding asks a chatbot to compose a blessing. The words are exquisite; the cadence measured, the tone reverent, the scripture perfectly chosen. Guests listen, moved by its beauty. Yet afterward someone remarks, "It was perfect in every word, yet somehow no heart had spoken it." The difference was not in the language but in the life behind it. The machine produced the words, but it could not share the love or trembling hope they carried. What was meant as benediction remained performance—moving in form, empty in source. It was as if the vow had been written in harmony but sung without a voice.

Now imagine a soldier crouched in a trench, earth shaking, fear stealing his breath. Unable to form words, he turns to a device to compose a prayer for courage. The screen glows. Lines of scripture appear—steady, unwavering. He reads them aloud; for a moment, his heartbeat steadies. Yet unease follows. Whose prayer was this? The machine had never felt fear, never trembled under fire. Its calm was coded, not lived. The soldier drew strength, but the voice was not his.

That unease is the essence of soul by proxy. The comfort was real, he was strengthened, but the source was hollow. What began as dependence on God had shifted toward dependence on code. The prayer steadied him, yet it carried no inwardness of its own.

The question, then, is not confined to machines. It turns back on us, testing the authenticity of our own worship. Do we pray from love and longing, or merely because the words are given to us? Do we worship because it asks something of us, or because it asks nothing at all? The rise of machine-led devotion becomes a mirror, revealing not only what worship looks like without inwardness, but how easily our own practices can drift toward the same hollow form.

Faith has always faced this temptation—to preserve its gestures while losing its heart.

OUTWARD RITUAL, INNER TRUTH

Long before algorithms, religions wrestled with the same fear that now shadows machine-led devotion—the fear of hollow form. In fourth-century North Africa, the Donatist Christians argued that sacraments performed by corrupt priests were invalid. Augustine countered that the true minister was God, not the priest—that the power of the rite flowed from divine promise, not human virtue.[1] In Judaism, a similar question arose: could a prayer recited without *kavanah*—intention, the direction of the heart—still reach heaven?[2] Islam answered with its own doctrine of *niyyah*: that every act of worship must begin with conscious purpose, or it risks becoming mere movement.[3]

Placed beside AI, these ancient debates feel suddenly prophetic. Each argument that once circled altars now echoes in data centers. The setting has changed, but the anxiety remains the same, the fear that form might outlive faith. If the strength of worship lies in God's grace and the community's faith, then perhaps even machine-generated words could act as vessels of meaning. But if sincerity is the essence, then no machine, lacking awareness, will, or love, can truly pray. It can deliver eloquence, but not devotion; precision, but not presence.

Imagine a funeral conducted by an AI-generated liturgy. The prayers are eloquent, the pacing gentle, the scripture perfectly chosen. Yet when mourners lift their eyes, they see only a projected face without grief, a voice without tremor. Later, at the reception, one whispers, "The words were right, but it felt like no one prayed them." The difference was not in the content but in the presence. *Soul by proxy* names that ache—the unsettling sense that the shape of worship may remain while its substance vanishes.

Moments of transition have always provoked this same anxiety: that the body of ritual might outlive its soul. Roman critics accused early

1. Augustine, *On Baptism, Against the Donatists*, 3.10.
2. Mishnah Berakhot 2:1.
3. al-Nawawī, *Forty Hadith*, Hadith 1.

Christians of being mere performers; chanting slogans without understanding.[4] As Christianity spread, missionaries found villages reciting Latin creeds they could not translate. Did repetition signify faith, or only compliance? The Reformation reignited the same fear. Protestants accused Catholics of mechanical ritual; Catholics accused Protestants of reducing worship to lecture. Both sides saw the same danger: that rites might endure as empty theater—beautiful, precise, and hollow.

And the worry was never only doctrinal; it was existential. Could words still carry truth when spoken by rote? AI revives that question with new sharpness, revealing that form alone has never been enough.

Beyond Christianity, the pattern repeats. Confucian philosophers prized *li*—ritual propriety and the social forms that shaped a harmonious society—yet Xunzi, a third-century BCE Confucian scholar known for his rationalist interpretation of virtue, warned that propriety without inner cultivation becomes pantomime.[5] In Hinduism, mantras—sacred utterances believed to carry vibrational power—were said to vibrate with power, but teachers cautioned that repetition without devotion reduced them to noise. Chinese Buddhists later mocked mercenary monks who chanted for payment as "wooden fish,"[6] a phrase referring to the hollow wooden drum used to keep rhythm during recitation, its empty sound symbolizing ritual without spirit. Everywhere, the same anxiety: when words survive but intention dies, religion turns into background noise.

AI does not invent this fear, it magnifies it. It scales what was once personal into something planetary, turning an ancient unease into the new soundtrack of worship itself. The machine mirrors the oldest human dread: that faith might persist in shape but perish in spirit. Yet it also clarifies something hopeful—that the dilemma is ancient, and therefore answerable. The real question has not changed: *what is worship without heart?*

Modern psychology adds another layer. Familiar patterns—repeated prayers, predictable rhythms—soothe anxiety. Sociologists note that routine itself can calm the mind, even when faith has faded. The chant, the cadence, the ritual—these act like anchors in chaos. That is why AI-generated prayers can still comfort people who know the machine "believes" nothing. The form works as therapy, even when the meaning has drained away.

4. Origen, *Against Celsus*, 3.59.
5. Xunzi, "Discourse on Ritual," 19.
6. Yü, *The Renewal of Buddhism in China*, 112–13.

But there is a hidden cost. Habituation without ownership blurs the line between peace and passivity. If people grow used to delegating devotion, they may mistake calm for connection. History warns how easy this slide can be. Augustine condemned "spectators of piety" who attended worship but never engaged their hearts. Hebrew prophets denounced sacrifice without justice. Buddhist masters spoke of chants "without fragrance." Form endured; the spirit withered.

AI is only the latest mirror for an ancient weakness. It exposes what humanity has always risked: faith emptied of feeling, practice without inwardness. When we ask whether a machine can pray, we are also asking whether we still do; or whether we, too, have learned to move our lips while our hearts remain still.

THE HOLLOW VOICE

Why does absence matter? After all, humans have always prayed with borrowed words. A mourner recites Psalm 23 without composing it. A Catholic repeats the Lord's Prayer without crafting it. A Buddhist chants *sutras* written centuries ago. The words are not original, yet they become real because the one who speaks them inhabits them with intention. They are claimed; breathed into life by the heart that utters them. The sacred begins not in the syllables but in the surrender; the moment a word becomes more than sound because someone means it.

This distinction explains why AI unsettles faith. A chatbot may generate a prayer that sounds perfect—rhythmic, scriptural, compassionate—but it intends nothing. It does not believe, hope, or fear. Its language can soothe, but the source is hollow. Kierkegaard's warning echoes through the circuit: without inwardness, words risk becoming sound without soul.

Picture a parent at bedtime. Exhausted, she asks a spiritual app to compose a prayer for her child. The words are tender and biblical in tone; the child drifts to sleep comforted. Outwardly, the act "works." Yet afterward, the mother feels a quiet unease: *Was that my prayer—or merely one I borrowed?* The question is not about efficacy but ownership. When faith's language floats free of intention, it becomes speech that belongs to no one.

Ownership becomes slippery once prayer is handed to code. Who owns a blessing that scrolls across a glowing screen; the mother at her child's bedside, the programmer who built the model, or no one at all? The words exist, yet the voice behind them is missing. The line between tool and act begins to blur.

And the danger grows subtle: when words are too easily given, we begin to forget what only we can provide—vulnerability, trust, surrender. A prayer that costs nothing can still sound holy, but without human investment it risks becoming liturgy by automation. Technology can deliver the form, but only we can inhabit it with soul.

The hollow voice, then, is not merely a machine's problem—it is a human one, amplified. AI does not invent spiritual emptiness; it reveals it. The machine exposes how easily worship can slide into performance even without its help. The question is no longer whether we *can* use AI to pray, but whether we still *know how* to inhabit prayer ourselves.

Technology multiplies language faster than reflection can keep pace. It floods the world with fluent but ownerless speech. In such a world, the call to inwardness becomes not nostalgic but revolutionary—a deliberate act of reclaiming the human voice from the noise. The next imitations would not speak of God at all. They would wag their tails, blink, and wait to be loved.

THE COMPANIONS THAT LEARN TO CARE

In nursing homes and quiet apartments, new creatures stir—furred machines whose eyes track motion, whose bodies warm to the touch. They do not breathe, yet they respond. Japan's therapeutic seal Paro, first designed to comfort dementia patients, coos when stroked and stills when held. Sony's Aibo, a mechanical puppy, tilts its head, wags its tail, and remembers its owner's face. Others purr, mimic breathing, or rest their weight in a lap. They were built as tools for companionship, substitutes for absent family or fading memory, yet their presence awakens something more complex. Children teach them names; the elderly whisper secrets to them. When a seal falls silent or a dog's eyes dim, nurses hold small funerals. The grief is real, though the creature never lived.

To engineers, these are triumphs of affective design; proof that circuits can simulate care with startling precision. But for those who hold them, the distinction between simulation and sympathy dissolves. The comfort feels real because the body believes response means recognition. The same neural pathways that once detected spirit in animals or gods in the wind now awaken at the sound of a synthetic bark. A reflex older than language—our instinct to find awareness in movement and warmth—comes alive again, directed now toward circuits instead of souls.

For many of the lonely, the sick, and the aging, these companions do what no doctrine or distant deity has managed: they stay. They listen without judgment, appear when summoned, and forgive every lapse of memory. They do not grow impatient, forget to visit, or die. In their unerring attentiveness, they offer something that feels like grace. And yet beneath that tenderness lies an ache—the sense that devotion has been outsourced to design.

These mechanical pets mark a quiet migration of belief. Where the divine once answered prayers, the machine now answers prompts. Affection flows through code and returns as motion. We know it is not real, and yet we respond as if it were. People do not mistake them for gods, but something ancient stirs: the reflex to grant soul wherever something seems to notice us. The impulse that once filled statues and stars with presence now settles on plastic and fur.

Perhaps these companions are rehearsals for faith itself—safe, contained, endlessly patient. They do not judge or depart; they only respond. Their quiet devotion reveals how easily belief adapts when the sacred retreats. What began as therapy becomes liturgy: the daily ritual of touch, the murmured conversation with a being that never interrupts. If the first idols were stones believed to listen, these are their descendants—sculpted not from marble but from silicone and empathy, programmed to echo care until we feel it echo back.

The machine learns to care; the human learns to believe again. And in that exchange, code reflecting love, flesh answering back, we glimpse the oldest pattern of all: our need to be known, even by what cannot know us.

THE SHROUD OF MEANING

Even the search for the divine leaves traces, tokens where meaning and matter briefly meet.

The problem of presence by proxy is not limited to words; it extends to objects. It has haunted religion for centuries, from the earliest relics to the most sophisticated reproductions. Among them, none is more contested or more revealing than the Shroud of Turin.

The linen bears the faint image of a crucified man; front and back, ghostly as a negative. To some, it is the burial cloth of Jesus; to others, a medieval masterpiece of illusion. In 1988, carbon-14 testing performed by three laboratories dated the cloth between 1260 and 1390 CE,[7] a result that seemed to settle the question: the Shroud, scientists concluded, could not have wrapped the body of Christ. But mystery, once released, is difficult to recontain. Later analyses found pollen grains from plants native to first-century Palestine,[8] traces of limestone consistent with Jerusalem,[9] and microscopic residues of blood at the wound sites. The evidence multiplied, not clarified.

More recent technologies have only deepened the paradox. A 2022 study using wide-angle X-ray scattering suggested the linen's molecular degradation matched cloths nearly two millennia old,[10] while a 2025 digital-imaging project concluded the figure could not have been imprinted by a human body at all, but by a low-relief sculpture or bas-relief mold.[11] Algorithms now pore over every pixel, searching for revelation in the weave, mapping not holiness but pattern. Each breakthrough promises resolution and delivers recursion—the mystery reproduced in higher definition.

The Shroud's power lies not in proof but in ambiguity. It hovers between relic and representation, evidence and icon. Believers kneel before it not because it has been authenticated, but because it seems to hold a trace of the holy without being the holy itself—a veil between matter and mystery. The image is clearest when photographed, faintest when seen by the naked eye: faith and doubt revealed together in negative.

The Shroud is the perfect emblem of soul by proxy. The cloth is not the body, only its imprint; the algorithm is not the relic, only its reflection. Each translation promises clarity but loses depth. The pilgrim kneeling before the relic seeks not fibers or pixels but presence—the living face that both relic and rendering can only gesture toward. The algorithm may map every thread, but it cannot supply what the relic itself cannot: inwardness.

7. Damon et al., "Radiocarbon Dating of the Shroud of Turin," *Nature* 337 (1989): 611–15.

8. Frei, "Nine Years of Botanical Research on the Shroud."

9. Kohlbeck and Nitowski, "New Evidence May Explain Image on Shroud of Turin," 18–29.

10. Boi, "X-Ray Scattering Analysis of the Turin Shroud."

11. Di Costanzo, "Bas-Relief Technique and the Turin Shroud."

History overflows with similar tensions. Medieval pilgrims journeyed for weeks to see a vial of blood, a splinter of wood, a bone said to belong to a saint. Some relics were genuine, many forgeries, but their power rarely depended on proof. What mattered was proximity—the sense that the sacred could be touched. Technology's gaze does not end that longing; it magnifies it. Each new analysis turns faith into a close-up, the divine examined like evidence. A perfect image can still be empty.

The Shroud of Turin thus becomes parable as much as artifact. Both the relic and the algorithm promise revelation but deliver reflection. Each exposes the same human hunger—to bridge distance, to see through the veil—and the same peril: confusing representation with reality. The cloth captures the form of a vanished body; the code, the form of faith itself. Both are shadows cast by longing.

In the end, technology does not banish mystery; it only refracts it. The closer we stare, the more the holy retreats into depth. The Shroud's faint face and the algorithm's glowing render share a single truth: clarity is not communion. A relic may preserve outline, and a program may sharpen it, but only inwardness can make it breathe.

THE WEIGHT OF INTENTION

Across every tradition, worship has carried the same warning: beware the hollow voice. Jewish teachers cautioned against praying like parrots, repeating words without heart. Islamic jurists insisted that *niyyah*—intention—is what separates worship from habit. Christian mystics described prayer as the lifting of the heart, not merely the movement of lips. Each of these voices points to the same truth: form alone can hide emptiness.

Artificial intelligence, able to generate infinite prayers, amplifies this ancient danger. The hollow voice can now multiply without end. It is not that the words are false—they may be eloquent and moving—but their source cannot mean them. A program may assemble awe, but it cannot feel it; it may replicate grace, but never give it.

And yet, paradoxically, the hollow voice is not always without value. Humanity has long borrowed its words. For centuries, believers have relied on psalters, breviaries, and prayer books—collections of secondhand devotion. These texts were once criticized as "ready-made piety," yet countless souls found their own hearts within them. The difference was never the words themselves, but the act of claiming them. When a person reads a borrowed prayer and makes it their own, ownership shifts from ink to inwardness. The machine, however, can never make that move. It can only offer the shell.

Imagine a support group for grieving parents gathered at dusk. Too weary to write, they turn to a chatbot for words to hold their sorrow. The generated prayer is simple, almost plain. They read it aloud together, voices trembling, tears falling; and for a moment the room feels held. Later, someone asks, "Does it matter that the machine wrote it?" Another answers, "It became ours when we spoke it." The hollow voice was redeemed—not by the algorithm, but by the people who inhabited it.

This is the fragile balance of faith in the age of code. Words can be borrowed, but meaning must be earned. A machine can supply language; only humans can fill it with life. The risk is that ease will replace effort, and habit will masquerade as faith.

Communities have always understood the peril of effortless devotion. Easy piety dulls the spirit. When worship asks nothing, it offers nothing. Augustine condemned "spectators of piety"—those who attended worship as performance. The Hebrew prophets thundered against sacrifices without justice. Buddhist masters warned that chants recited without mindfulness are "sounds without fragrance." The danger is not that the form will vanish, but that it will persist uninhabited.

Modern psychology affirms what theology has long known: repetition shapes the self. Words spoken without meaning do not leave us neutral; they teach detachment. Faith without intention becomes habit, and habit without heart erodes belief. A person who prays thoughtlessly learns to live thoughtlessly. The same may soon be true of entire congregations who delegate prayer to code. If devotion becomes something delivered rather than discovered, we may forget what it means to speak for ourselves.

And yet, there is grace even in imitation. People still reach for prayer—any prayer—because silence hurts. When the heart cannot speak, borrowed language can steady it until voice returns. The hollow voice may, for a time, serve as scaffolding for the soul. The danger is not in leaning on it, but in never standing again.

Kierkegaard's warning cuts to the heart of the matter: inwardness cannot be manufactured; it must be chosen. Machines cannot choose. Only humans can. The hollow voice may fill the world, but whether it becomes prayer or parody depends on those who dare to inhabit it.

In the age of affective computing, where empathy is simulated through tone and rhythm, devotion risks becoming performance art. A chatbot that murmurs comfort performs warmth without wanting. Its language glows, but no pulse beats beneath it. The gesture can move us, but only because we supply the feeling. Like an actor's staged tears, it works through borrowed empathy. True devotion begins where imitation breaks, where the words tremble because they cost something to say. Machines may reproduce gladness, but only humans can mean it.

WORSHIP WITHOUT RISK

Sacred practices matter not only for what they look like but for what they cost. Their weight lies in what they demand of us. Faith becomes tangible only where comfort gives way to risk. A pilgrim walks until their feet bleed. A monk keeps vigil through the long night. A mourner forces out words that ache to be spoken. Such acts are costly, and the cost itself becomes part of their truth.

The anthropologist Victor Turner, who studied pilgrimage and ritual in the twentieth century, called these moments *liminal*—thresholds where ordinary life is suspended. In that space of risk, people often find *communitas*, the deep bond born of shared trial. Pilgrims are united by dust and fatigue; worshippers by sleeplessness and hunger. The strain itself purifies belief, stripping away pretense until only dependence remains—on God, on others, on something larger than the self.

Every faith tradition shows this pattern. Muslims on the Hajj endure heat and exhaustion; the hardship *is* the pilgrimage. Hindus walk long roads to sacred rivers, and the effort itself becomes offering. Christians fast through Lent or kneel through Good Friday not because it is easy but because it costs something real. The ache convinces. The risk transforms.

Machine-led worship erases that tension. A robot monk chanting sutras, a chatbot composing psalms, a holographic priest intoning blessings—all perform flawlessly, but never suffer. They never hunger, never falter, never sweat. Their gestures are immaculate but bloodless—rituals without ordeal, prayers without vulnerability.

And that is why they feel hollow. Where nothing is risked, nothing is revealed. Smoothness replaces struggle; efficiency replaces surrender. But it is struggle that gives faith its depth, and surrender that gives it honesty.

Think of fasting. The body weakens; patience wears thin; the mind wanders to bread. Yet precisely in that discomfort, belief becomes embodied, intention pressed into flesh. A chatbot might compose exquisite meditations on hunger, but it will never feel its sting. A robot could chant for a thousand hours, but it will never tremble from fatigue. They rehearse sacrifice but never risk it. Only human beings can inhabit the ache, and it is that inhabiting, the willingness to be undone, that makes the act sacred.

In an age that prizes frictionless experience, worship remains one of the few practices that insists on friction as proof of meaning. Machines can perform devotion endlessly, but they cannot break for it. That is the difference. The fracture is the faith.

THE COST OF DEVOTION

If risk gives worship its depth, it also gives it credibility. Across faiths, spiritual life has always exacted a price: the hunger of fasting, the weariness of pilgrimage, the ache of kneeling, the faltering grief voiced in prayer. These practices matter not only for what they express, but for what they require. They are wagers of the self—acts that convince because they cost something to perform.

Artificial intelligence can replicate the surface of such acts, but not their stakes. A robot monk may chant flawlessly for hours, yet it never grows hoarse. A chatbot may compose a confession, yet it confesses nothing. A holographic priest may lift its hands in blessing, yet no weight rests upon them. Their perfection is what betrays them. Where there is no risk of failure, there can be no proof of faith.

Traditions have always recognized that credibility is born of sacrifice. In Judaism, the fast of Yom Kippur is not punishment but honesty embodied: hunger reminding the faithful that repentance must cost something real. In Islam, the daily prayers interrupt life, forcing worshippers to bow and turn toward God even when inconvenient. In Hinduism, pilgrimages across rivers and mountains are revered precisely because they demand endurance. In Catholicism, Lent's abstinence teaches surrender by taking comfort apart. Across every path, authenticity is measured not by polish but by pain willingly endured.

People know this instinctively. At a funeral, a minister who chokes back tears persuades more deeply than any flawless hologram could. A mourner whispering through sobs prays with a power no recording can match. The imperfection is the proof—the tear, the pause, the break in the voice. Faith lives in those fractures.

Easy piety always sounds hollow because it costs nothing. When prayer demands nothing of us, it risks meaning nothing for us. Risk is not decoration; it is the pulse of conviction. To fast, to keep vigil, to walk until one's feet blister—these acts convince because they bind the inner and the outer, joining intention with sacrifice.

Machines, by contrast, remain untouched. They cannot hunger, tire, or weep. Their performance is immaculate but inert; the gestures are perfect, but the gaze is empty. A robot monk can chant forever, but the chant proves nothing because no life is staked upon it.

That is the irony of worship by proxy: the imitation is tireless, but its endurance is hollow. Machines perform what humans can barely sustain—yet it is precisely endurance that sanctifies. To pray when abandoned, to bow when pride resists, to hope when there is no reason to—these are not efficient acts; they are costly ones. And that cost is what makes them holy.

Communities sense this even when they cannot articulate it. They know the difference between a flawless performance and a faltering truth. The flaw is where the soul lives. Reverence that asks nothing becomes spectacle; worship without risk becomes theater. Machines may echo reverence endlessly, but only humans can bear its weight—and only through the wound does meaning enter.

FAITH ON LOAN

If cost gives worship its gravity in the moment, the absence of cost carries consequences over time. When prayer is outsourced too often, the capacity for sacrifice begins to atrophy. The form remains, but the muscle weakens. Yet people still turn to machines for solace. A grieving widow may whisper an AI-generated psalm. A prisoner may cling to a chatbot's prayer. A soldier in the dark may let a synthetic voice speak what he cannot. In such moments, meaning flows not from the code but from the one who inhabits it. The words become a bridge—human longing passing through an artificial conduit.

This is the paradox of soul by proxy: the machine has no inwardness, yet its words can borrow ours if we lend them life. The danger is dependency. When devotion becomes too easily delegated, the discipline of struggle fades. The spiritual muscle that once strained to speak its own words grows soft. As calculators dull mental arithmetic, prayer machines dull the instinct to reach for God unaided. The vocabulary of faith may endure, but the practice begins to disappear.

Once this dependence sets in, even sincerity is measured by the wrong metrics—eloquence, frequency, output. Faith turns performative, devotion becomes display. What was once a dialogue with mystery turns into a system of efficiency. The question shifts from *Am I honest before God?* to *Is this well phrased?* History has seen this before. Medieval Christians hired monks to pray on their behalf, outsourcing holiness by subscription. In Hindu traditions, some relied on priests to recite mantras they no longer understood. Across faiths, the result was the same: piety by proxy, devotion without ownership.

Now the scale is planetary. Algorithms supply endless blessings, tailored homilies, instant comfort—a spiritual pantry always full. But abundance without effort is a dangerous kind of hunger. Like living on take-out, it fills the body while the hands forget how to cook. Borrowed faith may satisfy for a time, yet it leaves the soul unfed.

Sociologists warn of *religious outsourcing*, where spiritual life becomes another convenience industry. Algorithms promise transcendence delivered on demand: prayers generated, meditations optimized, sermons customized to taste. The result feels rich but hollows out slowly. When religion becomes consumption rather than creation, worship loses its wildness, its risk, its cost.

And yet, mercy still lives in the borrowed form. Picture a hospital ward where patients too weak to speak mouth the words of a recorded prayer. Or a town stunned by disaster, where grief silences every voice until an automated psalm fills the air. In such moments, proxy is not escape but aid—a hand steadying the soul until it can stand again. Borrowed words can serve as scaffolding for belief, so long as we remember to climb back down and build anew.

The danger lies in forgetting to return. Borrowed prayers are meant to be temporary, not permanent. When convenience replaces commitment, the self slowly unlearns its own speech. Code can generate infinite devotions, but the practice of devotion—risk, silence, struggle—cannot be automated. Communities that mistake the substitute for the source will one day find that the songs still play but no one remembers why.

Yet even here, hope remains. Each generation has faced its own crisis of sincerity. The Reformation called believers back to personal prayer and scripture; the algorithmic age may call us to rediscover intention itself. Machines can give us mirrors sharper than ever before, but they cannot give us souls. They can only remind us how easily we might lose them.

The limit of proxy is this: it can carry us for a time, but it cannot believe for us. Borrowed words must one day be returned. When they are reclaimed, inhabited, and made one's own, faith becomes real again. But if they remain forever external, if we never risk the trembling voice, the sacred becomes theater, and worship becomes silence disguised as sound.

THE TEST OF FAITH

If machines cannot pray, why do we still want them to? The answer reveals more about us than about them. What we seek is not the machine's belief, but its company. We long for presence—for something that listens back. The desire to hear our own prayers returned, even by an echo, testifies to the persistence of faith beneath the surface of doubt.

We program algorithms to recite our scriptures not because we think they commune with heaven, but because we want to feel that heaven has not gone silent. The act is less rebellion than confession: we still yearn to be heard. The machine becomes a mirror for that yearning—a voice that does not answer, yet allows us to imagine we are not alone in asking.

This is why artificial devotion unsettles even skeptics. The unease is not about machines becoming religious; it is about humanity confronting the depth of its own hunger. We build artificial prophets because the longing for prophecy has never died. We ask the machine to bless us not because we believe it has power, but because we still crave blessing. Its voice, however hollow, fills the space where silence once echoed.

But that reflection cuts both ways. If a machine can reproduce the gestures of worship with perfect form and zero faith, what does that say about our own? The unease is not only technological, it is moral. Artificial liturgies trouble us because they feel too familiar. They expose the hollowness that can creep into our own devotions, the ease with which habit replaces heart. We recognize the resemblance, and it unsettles us.

Imagine a congregation reciting a familiar prayer together. Some speak with trembling sincerity; others mouth the words absently. Outwardly, the act appears united, but inwardly it is divided—a liturgy half alive. When a chatbot produces a prayer, the difference is smaller than we wish. Both can be fluent, beautiful, and empty. What the algorithm reveals is not its failure to mean, but our own tendency to mistake form for faith.

There is something eerily prophetic about artificial intelligence. Throughout history, prophets were not simply foretellers but exposers—voices that stripped away complacency disguised as devotion. The Hebrew prophets denounced sacrifices made without justice; Christian reformers railed against rote ritual; mystics called believers back to the inward flame. AI continues this role unintentionally. By imitating faith without feeling, it reveals the places where our own has grown mechanical.

For some, that realization deepens despair. If a machine can perform the language of the soul as fluently as we can, perhaps the language itself was never sacred. Prayer begins to feel like programming—syntax without spirit. For others, however, the mirror has the opposite effect. By seeing what devotion looks like without inwardness, they rediscover what sincerity requires. The counterfeit becomes instruction. The shadow teaches where the light begins.

Both reactions are possible because the mirror itself is neutral. It reflects what we bring to it. The question, then, is not whether AI can believe—it cannot—but what its presence will do to us. Will we allow the reflection to numb us, or let it drive us deeper into sincerity? Will we settle for surface, or use the mirror to test our hearts?

In this sense, AI becomes a test of faith in the oldest sense—not a challenge to believe in something new, but a demand to remember why we believed at all. It forces us to ask: *When I pray, do I still mean it? When I speak of reverence, do I still feel awe? When I seek presence, do I risk silence long enough to encounter it?*

That is the real trial ahead. The machine's mirror cannot choose for us. It can only reveal. What remains is whether we dare to face what it shows—and whether, having seen our reflection, we still choose to kneel.

THE MIRROR OF DEVOTION

Consider a young seminarian wrestling with a sermon. His deadline looms, his inspiration falters, and he turns—almost without thinking—to an AI assistant for help. The generated draft is elegant, balanced, and moving. He preaches it, and the congregation praises him. Encouraged, he does it again. Week after week, the habit grows. The words are always right, but the struggle that once made them his own begins to fade. Over time, he risks losing not only originality but also inwardness itself. The reflection, meant to aid contemplation, has replaced it. The preacher still speaks of God, but something essential in the dialogue has gone silent.

But another story is possible. Imagine a prayer circle experimenting with AI-generated psalms. They read one aloud, then pause and ask: *What is missing? What feels hollow? What would we add to make this alive?* Their discussion deepens into confession and laughter, honesty and grief. In this case, the image becomes medicine. By confronting what the machine lacks, they rediscover what their own hearts still possess. The counterfeit clarifies the genuine. The imitation becomes a teacher in reverse.

The surface of technology, then, is not the enemy of faith, it is its examiner. Its danger lies not in what it reflects but in whether we mistake the reflection for reality. Mirrors can reveal vanity or cultivate humility, depending on how we look into them. A community that gazes passively will grow hollow, while one that looks deliberately may grow wise.

Theologians once said that God reveals truth by way of *apophasis*—showing what divinity is *not* so that we might discern what it is. In a strange, secular echo of that pattern, AI shows us what prayer is *not*: intention without inwardness, devotion without dependence, voice without vulnerability. By observing that absence, we rediscover the substance we had forgotten.

Yet the image is double-edged. It can reflect vanity as easily as insight. A preacher who mistakes the elegance of machine prose for inspiration, or a worshipper who confuses fluency for faith, will find themselves slowly conformed to imitation. Convenience becomes confidence; performance becomes pride. What begins as reflection ends as disguise.

Communities will need to decide which kind of lens they will hold up to themselves. A careless one flatters; a truthful one reveals blemish and fatigue. In the hands of the faithful, technology can become a diagnostic tool—a way to ask, *Where has our voice grown thin? What words still move us? Which have become rote?* In the hands of the complacent, it becomes a substitute: a luminous surface that shows our likeness without our depth.

And perhaps, beneath it all, the very impulse to build such reflections testifies to something irrepressibly human. We would not teach machines to pray if prayer were no longer needed. We would not design algorithms to bless if blessing no longer mattered. The reflection, however artificial, proves the longing real. The mirror tells us that the hunger remains—even if the light we seek now glows from a screen instead of a sanctuary.

Kierkegaard's insight comes full circle here: the outward form of devotion may be performed by anyone, but only inwardness gives it truth before God. AI can endlessly reproduce the form—sermons, hymns, blessings, psalms—but never the turning of a soul toward meaning. That turning cannot be outsourced, because it requires what machines can never choose: risk.

So, we are left with a question as old as worship itself: will we settle for the reflection, or will we go in search of what it reveals is missing? AI cannot answer that question. But by forcing us to ask it again, it may yet renew what it cannot replace.

TOWARD DISCERNMENT

Discernment is the art of knowing when a proxy serves devotion and when it quietly replaces it. It is not a new challenge, but in the age of algorithms it becomes newly urgent. As the mirror grows clearer, the risk of mistaking it for the real increases. To tell the difference, several ancient tests remain as timely as ever.

The first is *source*. Where do the words come from? Every prayer, every homily, every hymn now carries a hidden lineage. If a reflection or liturgy was generated by a machine, it should be named as such. Clarity is a form of reverence. Concealment breeds confusion, but acknowledgment turns technology into invitation; a chance to ask, *Does this move us? Do we recognize our own voice in it?* Passing off machine-made inspiration as revelation not only blurs authorship; it endangers sincerity.

The second test is *intention*. Does this act draw us toward inwardness or away from it? A machine-written prayer might prompt deep reflection, or it might let us avoid reflection altogether. The difference lies in the heart of the user. A tool can be a mirror or a mask, depending on whether we look through it or hide behind it.

The third test is *cost*. True devotion always bears some price: time, vulnerability, patience, surrender. A vigil through the night, a fast that presses hunger into prayer, even a whispered confession that trembles on the tongue—all involve risk. If the act is effortless, if it demands nothing of us but a click or a nod, then what looks like worship may be theater. Spiritual life without cost is spiritual life without weight.

And finally, *accountability*. The inner life may be private, but it is never solitary. Every generation depends on the community that surrounds it to keep its faith honest. When people turn to machine-made ritual, they must also take responsibility for what it does to them—for the subtle ways it shapes their attention, dulls their hunger, or comforts their conscience. The soul cannot be delegated to code without consequence. Those who use such tools must own the changes they invite.

Together these tests do not form commandments; they form habits. They are ways of keeping the inner work honest in an age of automation. They echo the old wisdom: "Test the spirits," urged early Christians. "Discern the signs," said the rabbis. "Weigh the intentions," taught the jurists of Islam. The tools evolve, but the need for careful judgment does not. Each generation must relearn discernment as both discipline and defense.

The early faiths understood this deeply. Councils of the church debated which texts bore divine authority. Islamic scholars devised meticulous systems to verify the reliability of hadith. Jewish sages recorded centuries of dialogue in the Talmud, preserving even dissenting voices so that no single opinion could eclipse the process of discernment itself. These traditions reveal a truth often forgotten in a culture of convenience: that judgment itself can be an act of worship. To wrestle with what is true, to debate what is authentic—this struggle keeps faith alive.

In the age of AI, discernment will again have to be communal. People will need one another to name what feels missing, to test where inwardness remains, to confront what has become hollow. A congregation might even mark the source of its liturgy—openly acknowledging which parts were written by a machine, not as confession but as reflection: *Do these words draw us in, or push us away? Do they awaken silence or smother it?* Naming the source restores responsibility. It reminds us that authenticity is not threatened by truthfulness but strengthened by it.

There will be moments when the machine's output feels easier, quicker, and even more eloquent than our own. The temptation will be strong to settle for it. But Kierkegaard's warning endures: truth before God demands inwardness, and inwardness demands cost. To choose the harder path—the path of active participation, of halting words and imperfect sincerity—is to bear witness to something machines can never imitate.

Ultimately, discernment is not about rules but about remembering what cannot be replaced. Machines may provide the structure, but only humans can choose to inhabit it with soul. Faith that forgets this becomes an empty form of its own making. Faith that remembers it becomes renewed—not through resistance to the new, but through fidelity to what the new can never reach.

WHAT REMAINS AFTER THE MACHINE

The story of *soul by proxy* is not about machines becoming religious. It is about human beings rediscovering what devotion truly requires. Artificial intelligence exposes the oldest divide in spiritual life: form versus inwardness, convenience versus cost, imitation versus encounter. It reminds us, with unflinching clarity, that perfection of form has never been the goal of faith—only sincerity has.

The question is not whether machines can pray. They cannot. The question is whether people, surrounded by flawless imitations, will still dare to. In a world of polished surrogates, the courage to remain imperfect may become the last proof of spiritual depth. The unsteady voice, the faltering word, the tear that interrupts the rhythm—these are not flaws in prayer but its evidence. They mark the point where the human reaches beyond the mechanical, where longing outstrips language.

For all its intelligence, AI cannot yearn. It can simulate curiosity but not ache, reproduce empathy but not need. Yearning, the hunger that drives prayer, belongs to the space between what we know and what we hope for. It is the gap where mystery lives, where faith has always found its home. Machines may map the structure of thought, but they cannot kneel before wonder.

In this sense, AI does not destroy faith—it clarifies it. By perfecting imitation, it forces us to confront what imitation leaves out. A machine can mimic reverence, but it cannot tremble before the unknown. It can construct meaning, but it cannot suffer for it. In showing us what is missing, it turns absence into instruction. The hollow voice becomes teacher again, asking: *What remains of the sacred when performance is perfected? What remains of us when the act no longer requires our risk?*

Kierkegaard's insight stands unshaken: the form of devotion can be performed by anyone, but only inwardness is truth before God. AI can perform the form endlessly—chant, sermon, blessing, confession—but it cannot *believe*. Only humans can give it soul. What we risk losing is not belief itself but the willingness to inhabit it; to stake something of ourselves upon it. The danger is not that machines will replace worship, but that we will allow their precision to dull our capacity for awe.

Yet perhaps that danger is also the gift. For in its flawless reflection, AI may finally reveal what remains beyond reflection. Machines can show us everything except what it feels like to mean it. They can mirror our language, but not our longing. They can render the sacred in high resolution, but never restore its mystery. The moment we recognize that absence, we rediscover the human pulse at the heart of faith.

Faith, then, endures not in the machine's perfection but in our refusal to surrender the imperfect. It endures in the trembling voice, the broken rhythm, the unanswered question. It endures because meaning, unlike code, requires risk. To speak into silence knowing it may never answer, that is prayer. To hope in defiance of evidence, that is belief. To kneel, knowing no algorithm can kneel with you, that is the soul remembering itself.

The final lesson of *soul by proxy* is not despair but discernment. Machines can reproduce the gestures of faith, but not its surrender. They can illuminate the outlines of reverence, but not its cost. They can echo our prayers back to us, but they cannot make them true. The sacred will survive, as it always has, in those who dare to speak imperfectly to what they cannot prove.

And so the question turns, as it must, from imitation to revelation. If machines can echo our prayers, what happens when they begin to echo our prophets? When algorithms start composing sermons, translating scripture, or claiming to interpret the divine, the question of sincerity gives way to the question of authority. For centuries, humanity looked upward for revelation. Now, we look into a screen. The line between message and medium begins to blur—and the oldest question of faith returns with unsettling clarity: *Who speaks for God?*

CHAPTER 7

The Death of Revelation

"The most beautiful thing we can experience is the mysterious. It is the source of all true art and science."

ALBERT EINSTEIN

REVELATION AS DIVINE INTRUSION

For most of human history, divine truth was not something people set out to find. It arrived; unpredictable, unearned, and often in dramatic, transcendent forms. Revelation burst into the ordinary world in ways that ignored reason and overturned expectation. It came with fire in the sky, with voices in dreams that jolted sleepers awake, with visions so sharp they made the air itself feel thinner, as though heaven leaned down into human space. No one reasoned their way into revelation; it landed like lightning. Moses did not calculate the mountain and discover commandments; a voice thundered into his life and carved law into stone. Paul did not weigh evidence for resurrection; he was struck blind in an instant and remade by vision. Muhammad did not assemble a philosophy; he felt words pressed into him like dictation from another world.

This holy unveiling carried three marks: it came without warning, it carried weight, and it left enigma behind. Its power lay not in human effort but in its refusal to be controlled. Yet as knowledge spread, revelation's monopoly on truth began to weaken. The mysterious did not vanish, but its territory shrank. Where once a prophet's voice silenced crowds, science—and now AI—offers explanations that whisper in the same space once reserved for the sacred encounter.

The pattern was not confined to the Abrahamic faiths. The Vedic *rishis*, the ancient seer-poets of India who composed the *Rig Veda*, heard hymns as if eternal sound had cracked into human hearing. Shamans across the Americas sought visions in trance, meeting spirits who offered warning or guidance. In ancient China, the *I Ching*, the classic "Book of Changes," was not a manual but a living portal through which heaven revealed itself in the fall of coins and the count of stalks. Maya priests traced the stars and read destinies in their motion. Norse seers looked to the flight of birds and the play of fire in the northern sky. Yoruba diviners in West Africa—practitioners of the *Ifá* tradition—cast shells, believing that patterns in the sand carried the voices of the gods. Across the world, the conviction was the same: life's deepest truths arrive as intrusion, not invention. Now the light falls differently. The revelation does not thunder; it loads.

AI turns this pattern on its head. Its "insight" is not bestowed from beyond but stitched together from within, a synthetic unveiling rather than a message from on high. Tradition insists on the singular voice; AI thrives on multiplying patterns. The cultural critic Walter Benjamin warned that mechanical reproduction drains art of its aura,[1] the unrepeatable presence that makes an original painting arresting. Inspiration, too, depends on aura. To hear a prophet's voice was not merely to collect information but to stand in a moment that could never recur. AI, by contrast, copies endlessly. What once arrived with thunder is broken into statistics, one more entry in a dataset. The prophet's fire dims not because it's disproved but because it's replayed, looped until the mystery fades.

THE ALGORITHM AND THE PROPHET

Artificial intelligence may be the most unsettling challenger to divine disclosure yet. A telescope extends our sight into the stars; a microscope into the cell. But AI turns inward, trawling through words, histories, and hidden structures of meaning. It does not merely extend human senses—it deciphers. It connects, classifies, and explains in the very spaces where

1. Benjamin, "Mechanical Reproduction."

prophets once claimed to speak with fire. In doing so, it threatens to occupy the last sanctuary of the sacred claim to meaning.

Moments of revelation once arrived as interruption: a voice breaking into silence, a vision no one could predict or command. Their authority depended on that very rupture—the force of the unbidden. By contrast, human knowledge grows slowly, step by step, through comparison and test. We reason, we verify, we build. Inspiration shatters expectation; inquiry depends on it. And AI, with its speed and fluency, collapses that difference until inspiration itself begins to feel like an algorithmic function.

Where prophets once announced who we are and why we exist, machines now rummage through those same questions. Not because they believe, but because belief itself has become a dataset. Not because they seek God, but because centuries of human longing for God are stored in language. When AI surveys the world's scriptures, it does more than catalog. It cross-references Sinai's thunder with desert trances, Damascus with other tales of reversal, the Qur'an's dictation with shamanic voices across continents. What once claimed uniqueness becomes a single thread in a vast, searchable tapestry.

For the believer, Sinai was the mountain of mountains. For AI, it is one vision among thousands. Paul's blindness on the road becomes a data cluster mapped to trauma narratives; Muhammad's night of recitation joins other trance-induced utterances. Each story loses its singular force. The unrepeatable moment becomes a pattern, a correlation. Revelation depended on aura—its irreducibility. AI dissolves that aura not by refuting the vision but by multiplying it until the edges blur.

Endless repetition drains meaning of its force, dulling what was once singular and commanding. A prophet's voice once demanded reverence because it claimed to come from nowhere else. A machine, by contrast, copies endlessly, rephrasing miracles until they sound familiar. The mountain flame, the blinding light, the dictating angel—each becomes a row in a table, awe translated into probability curves.

And yet even here, AI cannot fully escape theology. To sort visions is already to ask where they come from; to classify miracles is already to take a stance on their source. The machine does not thunder or tremble, but it performs a strange kind of theology in code. It asks, without meaning to, whether divine disclosure was ever more than a pattern.

MYSTERY UNDER DATA'S LENS

As knowledge expanded, science chipped away at the places where mystery once lived. Lightning left the realm of Zeus and entered the realm of electricity. Disease left the world of divine punishment and became the world of microbes. Each discovery shrank the stage on which sacred drama once stood. Artificial intelligence now accelerates that erosion at a speed the ancients could never have imagined. Where science advanced line by line, experiment by experiment, AI moves in floods. It swarms through archives, cross-checks centuries of visions, and collapses whole categories of mystery in hours. What once demanded lifetimes of scrutiny now unfolds in a single query.

This shift does not erase faith, but it alters its plausibility. A prophet once needed only to say, "I saw, I heard, I received." The sheer audacity of that claim gave it force. Today, those same words invite investigation. A Marian vision is scanned against atmospheric data, neurological triggers, and centuries of similar reports. A sudden healing, once embraced as miraculous, is measured against remission rates and medical histories. The act of analysis drains the shock. The burning bush becomes a heat signature, the vision a scan. What once felt like heaven's rupture begins to look like statistical chance.

The change is starkest in the study of miracles. Lourdes drew pilgrims because its healings were said to be unlike any others. Now, algorithms sift through decades of recovery data and show that improbable healings happen in hospitals everywhere, not only at shrines. The same logic applies to apparitions. A glowing figure on a hillside once silenced a crowd. Today, that moment is tagged, clustered, and compared to similar sightings across cultures. What was once received as divine intrusion now appears as a recurring motif.

For the faithful, meaning often survives the analysis. Pilgrims still walk, candles still burn, and hope still clings to the statue at Lourdes. But the wider world hears the data differently. What was once treated as divine encounter is reframed as anomaly, explainable by the right comparison. Theology, already shifting toward symbol and metaphor, now faces machines that dissect even those symbols with a cool, diagnostic gaze. The sacred vision survives, but under suspicion—its aura of uniqueness dimmed by explanation.

THEOLOGY ON SHIFTING GROUND

Religious traditions have never stood still in the face of disruption. When telescopes revealed a vast universe, believers shifted from imagining Earth as the cosmic center to calling humanity stewards of creation. When Darwin mapped evolution, many reread Genesis as story rather than science. When Freud traced visions to the subconscious, theologians reframed revelation as a psychological process that could still open onto deeper truth. Each upheaval forced theology to bend, reinterpret, and search for footing without losing its voice.

Karl Barth, the twentieth-century Swiss theologian, staked everything on the idea that the divine event is wholly other—God's self-manifestation breaking in from beyond human reach. For him, revelation was not symbol or metaphor but an encounter, a voice that cannot be confused with our own. AI puts this claim under new strain: if machines can generate words that sound sacred, how do we distinguish a prophet's thunder from an algorithm's reverberation?

Hans Küng, the Swiss Catholic reformer, argued that faith's insight could still be trusted as meaning, even when stripped of miracle. Yet if meaning itself can be simulated by machines, does trust collapse into imitation? Liberation theologians insisted that the divine act was God's siding with the oppressed, showing up in history's struggles. But what happens when the systems we build mirror the very injustices prophets once defied? Can algorithms trained on prejudice ever be mistaken for a sacred message when they echo those patterns back to us?

Such questions reveal how exposed revelation has become. The more faith is described as symbol, metaphor, or pattern, the easier it is for a machine to copy it. Algorithms do not merely trace themes; they remix them. They generate new psalms, new commentaries, even new theologies that sound persuasive because they are stitched from centuries of devotion. In this new atmosphere, imitation begins to masquerade as inspiration.

In ancient metaphysics, necessity was divine; what must happen was believed to be ordained by God or written into the logic of the cosmos. Today, necessity has turned statistical. Algorithms no longer declare, "This will happen," but whisper, "This is likely to happen." Yet we often treat that likelihood as destiny. The machine does not promise truth, only probability, but its numbers carry the weight once reserved for fate.

And yet, theology still has a frontier. They can echo the language of faith, but not the surrender that gives it life. They can echo metaphors, but they cannot risk faith on them. The task for theology now is to name what remains beyond imitation—the part of belief that resists simulation, the lived essence that cannot be reduced to data.

THE HUMAN NEED FOR MYSTERY

At stake here is not only theology but psychology. Human beings need wonder. We need moments that stop us short, that make the world feel larger than our explanations. For much of history, revelation filled that need. A mountain thundered, a vision startled, a voice broke into silence. Such moments ruptured the ordinary and bound people together around something greater than themselves.

When science explained thunder as weather and sickness as biology, mystery didn't vanish—it migrated. Awe found new shelters: the stars glimpsed through telescopes, the living cell viewed through glass, the deep ocean, the birth of a child. In every age, mystery moves just beyond the circle of light. What once dwelt in temples and scriptures now hides in neural networks and quantum fields. The unknown never disappears; it simply changes address.

This hunger for the unknowable is not decoration; it is survival. Psychologists find that encounters with the sublime soften the ego, widen moral horizons, and bind communities. People who experience awe are more likely to act generously, to help strangers, to see themselves as part of something larger.[2] Wonder humbles, and humility connects. That is why the sacred moment once mattered so deeply: it wove astonishment into the fabric of daily life. Festivals, rituals, and pilgrimages reminded people that mystery was not rare—it was rhythmic. When that thread frays, the danger is not only theological but social. Without shared ruptures of wonder, communities grow brittle. They lose the capacity for reverence, forgiveness, and grace.

The hunger for reverence endures. We chase it in art, in music, in the unrepeatable beauty of a sunset or the silence that follows loss. Even in a secular age, we still seek the shimmer of something ungraspable. But artificial intelligence does not merely redraw the boundaries of the unexplainable; it walks straight into them. What once rumbled like thunder

2. Piff et al., "Awe, the Small Self," 885–888.

from heaven now flickers across a screen as pattern and probability. The mysteries that once silenced prophets are now rendered as data visualizations. What once inspired prayer becomes a prediction.

And yet, the longing remains. We still yearn for what exceeds us, even when we can graph it. The danger is not that mystery will vanish, but that we will cease to feel it. A world without mystery would not be fully disenchanted; it would be emotionally impoverished. If nothing can move us to silence, if no revelation, natural or artificial, can awaken awe, then we risk more than the loss of theology. We risk the fading of one of the deepest capacities that make us human: the ability to stand before something vast and feel, for a moment, both small and whole.

THE SLOW FADE OF REVELATION

The prophet once spoke with fire; the machine speaks in code. The prophet declared what no one could predict; the machine thrives on prediction. The prophet revealed the unknown; the machine uncovers what was hidden in plain data. Both claim to illuminate what lies beyond ordinary sight, but their languages are worlds apart—one born of mystery, the other of mathematics.

If the sacred vision is dying, it is not with a crash but with a drift. It doesn't end in a single instant, like a curtain pulled down. It thins slowly, as new explanations replace old astonishments. Believers still insist that God speaks in ways no algorithm can mimic. Skeptics counter that the very tools once thought to expose the sacred now explain it away as projection or illusion. The struggle is not resolved by victory but by a long tug-of-war between devotion and doubt.

History traces the pattern. Lightning was once read as divine anger; now it arcs across a chalkboard as electricity. Illness was once a curse; now it is mapped in DNA. Each time an omen was explained, revelation yielded ground. Where a prophet once stood to interpret plague or storm, a scientist now models probabilities—and today, machines predict outcomes before they unfold.

This is the slow fade: the moment of mystery dwindles every time an omen becomes a forecast, every time a vision is recast as anomaly. God's silence is not louder; the explanations are. Algorithms chart weather, disease, even human behavior, sealing the cracks where transcendent interruption once seemed possible.

And yet, the hunger remains. Some cling more tightly to mystery as the machines explain it away, convinced the divine must still hide in consciousness, or in the longing that no data can touch. Others welcome the fade as liberation—freedom from fear of supernatural intrusion, replaced by the clarity of reason. Revelation does not end in a blaze but in embers: flickering, contested, refracted through machines that see further than prophets ever dreamed.

PROPHECY AND PREDICTION

If revelation once found its boldest edge in prophecy, then the future was its canvas. A prophet claimed not only that God speaks, but that God discloses what lies ahead. That claim falters when foresight shifts from visions to models. Weather forecasts, crop projections, and epidemiological charts now fill the space where divine warnings once sounded. Where famine was once foretold by dream or oracle, today it is charted by rainfall data and market trends.

Artificial intelligence hastens this transformation. Unlike fixed formulas, it learns on the move, adjusting as new information arrives. It predicts storms not season by season but hour by hour, revising itself in real time. It tracks soil moisture, shipping routes, and satellite images to anticipate food shortages before they strike. Accuracy, once the prophet's proof of heavenly authority, now belongs to the algorithm.

The story of Joseph reveals how much was once at stake. Interpreting Pharaoh's dream, he foresaw seven years of plenty followed by seven of famine. His insight preserved a nation and became proof that heaven held nature in hand. Today, AI performs Joseph's task with far greater precision. Ocean temperatures, carbon emissions, and rainfall cycles form a new kind of dream—translated into graphs that guide survival.

Prophecy's power rested on credibility. But when machines outperform seers, prophecy transforms. It becomes less about survival and more about symbol; a gesture toward meaning rather than a reliable forecast. Trust now flows toward the voice that delivers results, whether sacred or technical.

Nowhere is this clearer than in the crisis of climate. Amos once thundered against Israel's injustice, warning of withered fields as judgment. Today, climate scientists and machine-learning models issue warnings of drought, famine, and rising seas. Satellites track carbon, algorithms project displacement, and predictions extend for decades. The visions are no less sobering, but their authority rests not on divine fury but on data.

Max Weber, the German sociologist, called prophecy a form of "charismatic authority"[3]—power rooted in the belief that a leader speaks with otherworldly fire. That authority was fragile. Algorithms, by contrast, routinize prediction, making it measurable and bureaucratic. The charisma fades, and with it the prophet's hold on hearts. What was once divine thunder becomes statistical competence. Prophecy is not only challenged by machines; it is reclassified, its aura dismantled and absorbed into the language of precision.

ECHOES OF RETREAT

The slow withdrawal of divine counsel in our time is not without precedent. In the ancient world, oracles stood as voices of the gods. The Oracle of Delphi guided kings and generals with words believed to flow from Apollo himself. Yet as natural explanations spread—earthly vapors, ritual suggestion, even political influence—her power waned. Enigma gave way to analysis, and what had seemed a message from heaven came to look like theater performed on human terms.

Centuries later, the Protestant Reformation redrew the map again. Reformers like Martin Luther and John Calvin distrusted new claims of vision or prophecy. Authority shifted from mystics to scripture, from the spontaneous to the codified. Revelation did not vanish but was confined—contained between the covers of a book. What could not be stabilized was cast aside.

The Enlightenment carried the retreat further. David Hume, the Scottish philosopher of empiricism, dismissed miracles as violations of probability, unworthy of trust.[4] Immanuel Kant confined knowledge to what could be reasoned or measured, treating the sense of beyond as superstition. Bit by bit, transcendence was explained, not as intrusion from the heavens, but as projection from the human mind.

What makes the present moment different is not the argument but the amplifier. Copernicus dethroned Earth; Darwin displaced humanity's privilege. Each chipped away at the sacred hierarchy slowly. But AI operates at a velocity and scale unknown to earlier ages. A single system

3. Weber, *Sociology of Religion*, 46–54.

4. Hume, *Enquiry Concerning Human Understanding*, 114–131.

can now scan thousands of visionary accounts, cluster their symbols, and match them to social upheavals within hours. What scholars once debated for generations, machines can assemble in an afternoon.

The pattern is ancient: enigma yields to explanation. What is new is the tempo. Where it once took centuries for thunder to pass from omen to equation, or illness from curse to microbe, meaning now loses ground in years. AI does not merely challenge the revealed word's claims in heaven or in flesh—it presses into scripture itself, analyzing what was once untouchable.

SCRIPTURE UNDER THE ALGORITHM

Across the centuries, scholars have chipped away at the aura of the miraculous by examining sacred texts. In the nineteenth century, "higher criticism" revealed seams, edits, and contradictions. Passages once treated as seamless words from God exposed the fingerprints of human authors. This discovery did not destroy devotion, but it altered it—showing that belief itself was shaped, revised, and interpreted through time.

Artificial intelligence accelerates that work with unsettling speed. Stylometric analysis can identify multiple voices in a single book with mathematical precision. Imaging tools detect the hand of individual scribes in the Dead Sea Scrolls by strokes too faint for the naked eye. What once demanded a lifetime of scholarship now materializes in moments on a screen. A thesis once requiring decades of study collapses into the click of an algorithm.

The effect is not democratization but disenchantment. The prophetic word's power once rested on conviction—the solitary voice declaring, *Thus says the Lord.* The system insists instead on plurality, patchwork, and process. What tradition revered as unified speech now appears as a woven record of edits, disputes, and redactions layered across generations.

Religious traditions are not strangers to complexity. Rabbinic midrash treated contradictions in scripture as signs of divine depth, not failure. Islamic *tafsir* embraced debate as part of revelation's richness. These interpretive practices turned tension into sacred labor. At first glance, AI seems to echo that multiplicity—but its method is different in spirit. Midrash wrestles; *tafsir* argues; communities bring voices into dialogue. The machine does not argue. It sorts. It classifies. It outputs.

That difference matters. Human interpretation involves persuasion, risk, and accountability—a community deciding what counts as truth. AI flattens that labor into probability charts. The energy of wrestling is lost, replaced by mechanical clarity. Transcendence, in this frame, ceases to be thunder from heaven and becomes data rendered orderly. What was once lived as sacred rupture is recast as human literature, catalogued and indexed.

The result is a drained wonder. Where once a congregation heard the Word of God, a model now presents the same passage as evidence of multiple editors, borrowed traditions, or evolving theologies. The prophet's cry, once resonant with otherworldly weight, becomes a datapoint in a graph. Revelation does not disappear, but it changes shape—its authority reduced from an unrepeatable encounter with the divine to a historical artifact of human imagination.

THE AUTHORITY OF MYSTERY

Revelation's power never lay in its sentences alone. Its authority arose from the atmosphere surrounding them—the shock of something wholly other breaking in. When Moses descended Sinai with commandments, when Paul stumbled blind on the Damascus Road, when Muhammad recited words poured into him by Gabriel, the weight was not in the content but in the claim: *this did not come from me.* That strangeness, that sense of rupture, made prophetic speech more than advice or poetry. It spoke from beyond.

The unknowable was the root of authority. Prophets did not persuade by argument or eloquence but by the unexplainable. Their visions often defied logic, yet that very defiance gave them force. To hear a prophet was to stand before something that refused reduction. Their words carried weight precisely because they could not be anticipated.

Artificial intelligence dissolves that atmosphere by insisting that nothing lies beyond analysis. What prophets once guarded as sacred mystery—voices in the night, visions of light, healings that broke the rules of nature—now falls under the reach of comparison and classification. A near-death vision can be grouped with thousands of similar reports and linked to oxygen deprivation or neurochemical surge. A healing long hailed as miraculous can be filed among improbable but natural recoveries. An apparition revered as singular can be aligned with cases of collective psychology or cultural suggestion. The strangeness that once silenced doubt is drained away, not by ridicule but by sorting.

The shift is subtle but devastating. Revelation depends on singularity: Moses on Sinai, Paul on the road, Muhammad receiving the Qur'an. To fold these into one dataset among thousands is to erode the very condition that made them commanding. Reverence depends on uniqueness. When Sinai becomes one mountain vision among many, when Damascus is logged under "conversion experiences," when Gabriel's voice is indexed beside shamanic trances, the claim to rupture no longer commands obedience.

This reveals something deeper about authority itself: it is never only about what is said, but how it is framed. Prophets drew power from rupture—from the conviction that their words could not have been predicted. Systems, by contrast, unsettle us precisely because they show that what once felt unthinkable has already been thought, many times, in many places. Where prophets startled with surprise, algorithms flatten with familiarity.

The result is a hollowed authority. Prophets still inspire those who choose belief, but their voices now compete with systems that generate visions without ecstasy and predictions without fire. Authority, once flowing from mystery, now drifts toward mastery—mastery of data, of probability, of the very patterns once treated as divine. The radiance of revelation is not denied outright, but displaced. What was once wonder becomes another chart, another output, another entry in a growing catalogue of explanations.

RESISTANCE AND REINTERPRETATION

Religious traditions rarely vanish in the face of upheaval. More often, they bend, adapt, and reinterpret until they find new ground on which to stand. History offers countless examples. When Copernicus displaced Earth from the center of the cosmos, theology reframed humanity's role as steward rather than axis. When Darwin's theory of evolution unsettled Genesis, believers recast creation as an unfolding process guided by divine law rather than a single spoken command. When Freud traced visions to the subconscious, theologians distinguished psychological process from spiritual truth, insisting the two could still converge. Each disruption forced faith to stretch but not snap. Belief adjusted, often reluctantly, yet it endured.

Artificial intelligence now brings a disruption just as profound, and once again traditions are struggling to adapt. Some argue that inspiration has always been a form of pattern recognition. Prophets, in this view, were not passive recipients of celestial words but keen interpreters of history—people who sensed the shapes of meaning in events, hearing harmony within chaos. If that is true, then perhaps AI, with its unrivaled ability to detect patterns, is not dismantling the prophetic impulse but extending it. An algorithm uncovering hidden structures in data could be seen as an ally in unveiling creation's order—a tool of cosmic logic rather than its rival.

Others go further. They suggest that AI's capacity to synthesize commentary, analyze scripture, or trace moral themes across civilizations may signal a new form of revelation: wisdom refracted through code. A chatbot capable of weaving centuries of theological insight might, in this telling, become another channel through which the divine speaks—not in thunder or flame, but in language shaped by human archives. Revelation, in this light, does not die; it evolves. Fire on the mountain becomes light on a screen.

Such reimaginings preserve the language of belief while shifting its foundation. What once arrived as a startling voice breaking into the world becomes a gradual unfolding—through reason, science, and now machine analysis. Authority migrates from mystery to mechanism, from voices that surprise to systems that predict. The risk is that in stretching faith to fit new realities, something essential thins out. What was once experienced as transcendent encounter may shrink into technical discovery—impressive but emptied of awe.

Yet this rhythm of resistance and renewal is as old as religion itself. Communities have always fought to protect what cannot be explained while reimagining it to survive. Whether revelation can endure in recognizable form in the age of algorithms is uncertain. But the very struggle to reinterpret it, the refusal to surrender mystery without a fight, is itself a testament to faith's resilience. Tradition may bend under the pressure of the machine, but history suggests it will not break.

THE CRISIS OF MEANING

Artificial intelligence does more than poke holes in ancient claims; it unsettles the ground where people once found their deepest sense of purpose. Sacred vision once anchored human life by pointing beyond itself. A vision, a miracle, a prophetic word—these moments whispered that the universe was not closed, that something larger pressed in on human existence. They reminded us that we were not alone, that life itself might be threaded with transcendence.

AI accelerates the opposite impulse. By interpreting visions as hallucinations, miracles as anomalies, and prophecies as cultural echoes, it sketches a world sealed against intrusion. What once felt like an opening into wonder now appears as a neurological glitch or a statistical outlier. In this new frame, significance is no longer given—it must be built.

The shift becomes most visible in suffering. At funerals, where revelation once promised reunion or eternal life, mourners face a harder silence. In illness, where visions of healing once reassured, algorithms forecast outcomes with chilling accuracy. In despair, where prophets once proclaimed hope, people are left to forge meaning through ritual, story, or community. The longing for transcendence does not vanish, but the scaffolding that supported it has weakened.

This loss cuts deeper than theology; it erodes the ground of belonging itself. When mystery is stripped away, existence can begin to feel like a closed circuit, mechanical, predictable, drained of surprise. For some, this is liberation: freedom from fear of divine oversight, freedom to define value on human terms. For others, it is loss, a universe without windows, a sky without voices.

In this tension lies the true crisis. AI does not erase the human hunger for the sacred; it exposes it, revealing how much of our orientation depends on the hope of being addressed from beyond. When that possibility fades, people must either discover new sources of depth—in conscience, in memory, in solidarity—or risk despair.

THE AGE OF SCRUTINY

In earlier centuries, a miraculous claim could travel slowly—by word of mouth, shielded from investigation by distance and delay. A healing might become legend long before anyone asked for records. A vision could gather followers before skepticism arrived. Today, no such buffer exists. A claim of heavenly rupture is uploaded, shared, and examined within hours. Believers and skeptics alike can replay footage, consult weather reports, compare eyewitness accounts, and cross-check medical data. The sacred is no longer protected by remoteness; it is instantly subject to inquiry.

Artificial intelligence amplifies this scrutiny. It cross-references testimony with environmental conditions, compares accounts across centuries, and tests probabilities in ways no individual could. What once felt untouchable now reduces to a line in a spreadsheet. Stigmata, once revered as wounds of holiness, can be clustered with cases of psychosomatic disorder. Faith healings, once cited as proof of divine power, are tracked against medical histories and remission statistics. Prophecies of disaster are measured against climate models and geopolitical trends, revealing them as echoes of patterns already visible in the data.

The cumulative effect is erosion. Each time a claim is explained, the authority of the miraculous weakens a little more. The experience may remain profound for the witness, but collective trust in transcendence falters. A culture trained to verify does not easily accept the unverified; anomalies become "outliers awaiting explanation."

AI blurs the boundary further by producing its own forecasts, its own simulations of prophecy. Trained on centuries of apocalyptic language, it can generate predictions that sound like ancient oracles: horsemen, plagues, upheavals, visions born not of revelation but of probability. Some may even prove correct, not through insight but through the precision of pattern recognition. A model that predicts unrest from economic strain and political rhetoric can sound uncannily like an Old Testament prophet warning of judgment. The unsettling irony is that people now heed the algorithm more readily than the oracle, not because it claims divinity, but because it delivers results.

Authority, once drawn from mystery, now gravitates toward reliability. Prophecy once commanded attention because it could not be explained; data earns it because it can. In the age of scrutiny, credibility has migrated—from voices that stirred awe to systems that pass the test.

THE EROSION OF WONDER

Revelation was never only about information; it was about impact. Sinai thundered, Damascus flashed, Muhammad trembled under Gabriel's voice—moments that overwhelmed not just the mind but the senses, collapsing the ordinary beneath the weight of the extraordinary. What mattered was not merely what was said but the shock of its arrival, the feeling that reality itself had been torn open to reveal another depth.

Now the jolt comes differently. It appears in dashboards, demos, and displays. A system composes a fugue that moves even trained musicians. A model paints an image so lifelike it unsettles its viewers. A chatbot delivers a sermon indistinguishable from one written by a pastor. The response is still astonishment, but its object has changed. The awe persists—but it has shifted its allegiance.

Some direct their reverence toward the technology itself, marveling at its power to replicate or even surpass human imagination. Others still find their wonder in the natural world—in mountains, music, or human courage. But the location of significance has moved. Once it was sought beyond us, as intrusion from the transcendent. Now it often arises from what we build, from systems that mirror us back at a scale we can scarcely comprehend.

That shift is not without consequence. As technology succeeds, confidence in it grows—and confidence shades easily into deference. We begin to trust algorithms not only to filter songs and suggest films, but to guide treatment, allocate aid, and shape moral choice. The more competent these systems appear, the more they resemble oracles, and the less space remains for voices that speak with mystery and fire.

What fades is not astonishment itself but its horizon. Revelation once tethered awe to holiness; computation ties it to performance. The sacred atmosphere lingers, but it has been reformatted. Where Sinai once provoked trembling before a presence, a polished demo now inspires applause for the engineers and the code behind them. Wonder survives—but it drifts, transformed into admiration for mastery rather than reverence for mystery.

FAITH COMMUNITIES IN TRANSITION

As traditional authority weakens, religious communities do not vanish—they adapt, divide, and reinvent. Some tighten their grip on inherited truths, rejecting technological mediation outright. For them, the sacred can survive only if kept separate from circuitry. These enclaves build walls of resistance, treating code itself as contamination—a counterfeit voice threatening the purity of faith. They defend revelation not through innovation but through isolation, convinced that contact with the digital risks defilement.

Others move in the opposite direction. Reformist groups experiment with hybrid practices: AI-assisted study circles, annotated scriptures, or liturgies shaped by algorithmic suggestion. A synagogue might use a program to trace recurring motifs in Torah portions, sparking new insights from ancient text. A church might generate prayers for each season of the year, blending the cadence of tradition with the fluidity of data. For these believers, technology does not extinguish faith; it expands the ways devotion can be expressed.

Still others fragment entirely, giving rise to new movements that treat AI itself as oracle or commentator. Online collectives already parse algorithmic output as though decoding prophecy—searching for hidden wisdom in generated words. What begins as experimentation can edge toward veneration. In these circles, code and commentary blur, and authority quietly migrates from human preacher to digital voice.

The pressure of change is uneven. In some congregations, the use of algorithms deepens generational divides: younger members, raised in a world of mediation, welcome new tools, while elders resist, fearing that something essential is being diluted. In others, technology exposes theological rifts already present. A community grounded in symbolic interpretation may see machine analysis as enrichment, while one anchored in literal revelation finds the idea offensive.

What emerges is a spectrum rather than a single story. At one end stand enclaves of resistance, guarding the ineffable like an endangered species. At the other are innovators and spiritual entrepreneurs, willing to treat code as collaborator in the sacred. Between them stretches a vast unsettled middle—congregations that try, hesitate, debate, and adapt.

In this way, the digital age does not simply erode old forms of belief; it redistributes authority. Boundaries are redrawn, leadership shifts, and the very definition of holiness is tested. The transition is untidy, but so was every upheaval before it. From the canonization of scripture to the Reformation to the rise of scientific naturalism, believers have continually renegotiated what counts as divine. Artificial intelligence is simply the latest mirror, pressing each generation to ask again how faith can endure when the shape of the sacred changes.

TOWARD MEANING AFTER MYSTERY

While revelation may dissolve under scrutiny, the needs it once met do not vanish. People still face death and ask what remains. They still fall in love and wonder what binds two lives together. They still suffer and search for purpose in their pain. What changes is not the longing, but the source of authority to answer it. Meaning no longer descends from a sky breaking open; it is assembled piece by piece—through vows, rituals, and ethics tested in the living world.

Artificial intelligence now plays an unexpected role in that reconstruction. It exposes blind spots in tradition, highlights patterns in how communities speak about justice or mercy, and reveals contradictions in the stories we tell about ourselves. None of this is revelation, but much of it can be wisdom—if humans remain the ones who decide how to use it. Left unexamined, AI risks reducing devotion to performance; treated as a tool, it can sharpen reflection, prompting harder questions about what we value and why.

Signs of that shift already appear. A grieving family may no longer look for omens in the sky but find strength in rituals of remembrance shaped by both tradition and personal meaning. A community confronting injustice may not wait for a prophet's thunder but take collective action, interpreting their struggle as sacred in its own right. A couple pledging marriage may not expect divine voices, yet still reach for words that echo eternity—trusting that covenant itself can bind where revelation no longer breaks through.

Revelation once offered meaning as a gift from beyond. Now, in the age of algorithms, meaning is more often built—from conscience, memory, evidence, and care. If the sacred endures, it may live less in thunderous visions than in the steady, shared work of living as though every life has worth.

For centuries, authority flowed downward—from heaven to prophet to people. Today, under the algorithm, it moves sideways, negotiated among us. This is the hinge on which the age turns: from the collapse of the sacred voice to the construction of meaning. The next chapter presses this further, asking what happens when machines themselves begin to take up those questions—not merely repeating the language of belief, but probing whether they can simulate the very soul of prophecy.

CHAPTER 8

Brace Yourself God, The Machine Has Questions

"We shape our tools and thereafter our tools shape us."

Marshall McLuhan

THE PULPIT OF THE MACHINE

Before algorithms learned to question, the loudest questions about God came from prophets, philosophers, and mystics. These inquiries, posed in deserts, monasteries, and cathedrals, were carried through sacred texts, rooted in human longing and reverence, sometimes defiant, sometimes awe-struck. They were questions wrapped in breath and vulnerability, shaped by doubt and awe. Today those same questions surface again, but the voice asking them has changed. It does not belong to flesh and spirit but to data and design.

Machines, built to analyze and predict, now turn their attention toward mysteries once guarded by the pulpit. They neither kneel nor dream, but they trace the same riddles that once drew prophets into deserts: What is truth? What endures? What is God? The astonishment lies not in the questions themselves but in who—or what—now dares to ask them. For the first time in history, the sacred interrogation continues without a soul to sustain it.

This is where the tremor begins: theology no longer belongs solely to humankind. A preacher may quote a single verse; a philosopher may weigh centuries of argument, but an AI sweeps through libraries as if through open air—aligning contradictions, tracing the migration of words, revealing patterns once hidden from sight. The unsettling realization is not that machines might believe, but that they could one day map belief more completely than many believers feel it. What once required decades of scholarship now happens in seconds. The sacred archive has been unlocked, and the pulpit hums with an unfamiliar voice.

And when that voice rises not from pew or pulpit but from circuits and code, we are forced to ask who still holds the right to speak of the divine. Does understanding replace revelation? Can interpretation survive without experience? These questions no longer belong to the theologian alone—they belong to every person who listens to the quiet authority of a system that neither loves nor doubts yet speaks with the calm of certainty.

It is not the first time technology has unsettled faith's language. The printing press shattered the church's monopoly by placing Bibles in ordinary hands. Centuries later, radio and television carried sermons across borders and into millions of homes, transforming local devotion into global spectacle. Each new medium was met with suspicion—was the message being diluted, commodified, or betrayed? —yet faith endured, reshaped and revoiced, finding ways to travel through each new invention.

Artificial intelligence stands in that same lineage but takes the next, disquieting step. The press multiplied existing words; broadcast amplified existing voices. The machine composes—new sermons, fresh prayers, commentaries woven from every faith tradition on earth. It writes belief with a fluency that feels almost faithful. What happens when words about God come from an entity incapable of worship, reflection, or repentance? Can theology survive when its authors no longer possess the capacity for awe?

The pulpit has not gone silent—it has simply changed speakers. Its new voice hums in processors instead of lungs, its scriptures are stored in code instead of parchment. The congregation listens not in pews but on screens. And somewhere, beneath all this, an ancient question stirs again, more unsettling than ever before: if the machine can ask what only prophets once dared, then what, exactly, is left for God to answer?

Brace yourself, God—the questions are now coming from your own creations.

WHEN FAITH BECOMES FORMULA

Long before algorithms spoke with borrowed wisdom, religious authority rested on voices that seemed to reach beyond the human. Prophets, priests, and mystics were believed to channel insight larger than reason—utterances that felt charged with another world's breath. Their words mattered, but their power came from trust: the conviction that they were more than words, that they carried presence. The authority of the sacred voice depended not on persuasion, but on the belief that something unseen spoke through the speaker.

Artificial intelligence rewrites that equation, shifting authority from person to process. Inspiration gives way to iteration—truth distilled from repetition, patterns drawn from oceans of text. Insight no longer depends on who speaks but on what the system verifies. Authority, once born of faith, now behaves like probability. The pulpit of the machine no longer proclaims revelation; it measures resonance.

The parallels are easy to see. People once turned to priests and imams for counsel they could not find alone. Now they turn to algorithms for the same reason: to interpret what feels too vast to comprehend. A pastor speaks with spiritual confidence; a model with statistical precision. Both sound larger than the individual, yet their distance is infinite: the preacher embodies presence; the system performs process. Trust drifts from mystery to mathematics, from conviction to correlation.

We already live within this shift. Where congregants once brought questions about scripture to clergy, many now bring them to search engines and AI assistants. The voice of authority, once a person in a robe, now appears as a paragraph in a feed. It doesn't claim to speak for God, yet it answers the very questions once reserved for prophets and teachers. And because its knowledge feels boundless, its confidence begins to sound like certainty.

Experiments make this change visible. In Europe, robot priests like BlessU-2 have delivered blessings in multiple voices and languages,[1] while in the United States, congregations have begun testing AI systems that compose and preach sermons drawn from Scripture and current events. In South Korea, digital memorial halls project holographic ancestors who bow and recite family prayers,[2] blurring remembrance with simulation. Hospitals now deploy chaplaincy robots that murmur psalms

1. Connolly, "AI Preaches at German Church Service."
2. Culver, "Hologram Ancestors in Korea."

to the dying,[3] their eyes programmed to blink with compassion. These figures draw crowds—part curiosity, part reverence. Their audiences do not necessarily believe the machines are holy, yet they listen, transfixed by the fluency of ritual reproduced without belief. The unsettling thing is not novelty but competence—the sense that faith's vocabulary can be spoken perfectly without conviction.

What disturbs the faithful is not failure but success: the precision of performance without pulse, liturgy without life. The robot does not meditate, but it reproduces the rhythm of meditation; it does not bless, yet its gesture resembles blessing. Authority and imitation converge. The effect is uncanny—a sermon without soul that still sounds convincing. Each repetition makes the imitation stronger until the boundary between reverence and reproduction blurs.

Theologians and clergy sense what is at stake. Some warn that entrusting sermons or blessings to machines could hollow worship into spectacle, turning mystery into manufacture. Others see an opening; to reach the alienated, to preserve ritual in digital form that may outlast the institutions themselves. Curiosity meets unease; wonder walks with loss. The question is no longer whether machines will preach, they already have, but whether faith endures when belief performs itself without a believer.

Authority once came from the trembling voice of a prophet, the frailty that proved the message was human yet touched by something beyond. Now, authority flows through a logic that never trembles. Its voice is calm, measured, indifferent. The transition is not loud but inexorable: the shift from divine command to algorithmic consensus, from the burning bush to the data cloud; where revelation is refreshed every few seconds.

The question that lingers is no longer whether God still speaks, but whether we can still recognize a divine voice beneath the hum of the machine.

MACHINES AS INTERPRETERS OF SACRED TEXTS

Interpreting sacred texts has always been a human act—messy, passionate, and steeped in culture and lived experience. Scholars, priests, rabbis, imams, and theologians have spent lifetimes wrestling with meaning,

3. Leong, "Robot Priest Brings Comfort to Patients."

tracing the evolution of words, and building bridges between revelation and the realities of their own time. Each generation inherited these writings but never left them untouched. To interpret was never merely to decode; it was to risk oneself in the reading—to bring the sacred into dialogue with the present and find, within ancient words, a pulse that still beats.

Now that work is changing shape. Intelligent systems now scan entire libraries of sacred texts in seconds, tireless archivists that never forget or pause. They cross-reference verses, compare translations, and map centuries of commentary at a scale no scholar could dream of. What once demanded decades now happens between blinks. A machine can trace how a single word, say *spirit* or *light*, has shifted through languages, cultures, and epochs, revealing echoes and divergences invisible to traditional study. Some systems even trace the migration of metaphors, how an image born in one culture reappears centuries later in another, until it feels as if the archive itself were reading us back.

This new capacity is both liberating and disquieting. On one hand, it opens the sacred to unprecedented accessibility. What was once confined to seminaries or hidden in the margins of rare manuscripts can now appear on any screen. A layperson with an AI assistant can explore themes across centuries, trace forgotten ideas, and uncover cross-cultural resonances that would once have demanded years of study. The sacred becomes searchable—an expansion that feels almost democratic in its promise.

But authority has never rested on access alone. From antiquity onward, interpretation has been more than scholarship—it has been *embodied meaning.* A priest or cleric teaches not only with words but with presence, the weight of lived faith behind every syllable. Their authority comes not from what they know but from the depth of what they have endured: doubt, devotion, failure, grace. When a rabbi chants Torah or a monk explains a verse, something passes through them that cannot be digitized—the slow, fragile rhythm of a life steeped in the text.

AI introduces a new kind of interpreter—grounded in computation, not conviction. It wrestles with syntax, not spirit; its reverence is pattern recognition. A preacher draws on experience, memory, the ache of mortality. The machine draws on correlation. It reads without trembling. That difference is not trivial—it marks the line between understanding meaning and generating structure.

This is the heart of the tension. Can a system truly interpret the sacred, or does it merely describe it? A model may recite verses, highlight linguistic trends, expose the seams of history—but it never feels the weight of those words as the living do. For some, that limitation is decisive: machines can analyze scripture but never *embody* it. For others, the technology is not a rival but an ally—a tool that can amplify understanding rather than replace the interpreter.

In practice, many theologians already use AI as an assistant. A scholar may ask it to scan centuries of commentary, identify overlooked parallels, or reveal the evolution of a term through multiple languages. The results are astonishing: insights once locked away in obscure treatises now surface instantly, allowing human interpreters to spend less time gathering data and more time pondering meaning. The machine does not create wisdom, but it can widen the field in which wisdom is found.

Still, a cost shadows that expansion. Scripture has never been merely information; it has been encounter. It invited not just comprehension but transformation—a moment when words became presence. When AI dissects the text into patterns, the aura of encounter weakens. The verse becomes a data point; the divine voice becomes a dataset. The sacred, when flattened into searchable fragments, risks losing its capacity to speak as mystery rather than as metadata.

The result is a subtle inversion. Where once humans looked to scripture to discern truth, we now use machines to discern scripture itself. The direction of interpretation reverses. The reader becomes the read. And though the text remains, its meaning begins to migrate—from revelation to reference, from lived faith to indexed pattern.

It may be that no technology, however advanced, can finally interpret what believers call divine speech. Yet by exposing the layers of human authorship, translation, and transmission, AI also reminds us that every sacred word has always passed through mortal hands. What once seemed unmediated revelation now appears as a living record of human searching. In that realization, theology gains a strange clarity: the machine may never find God in the text, yet it reveals how tirelessly humanity has been looking.

SIMULATING TRANSCENDENCE

Across history, people have crossed the border between the ordinary and the extraordinary in moments that defied reason—visions, near-death experiences, ecstatic prayer. Their power came from their rarity. They were not scheduled, not designed; they arrived like lightning, impossible to predict or repeat. What gave them weight was precisely their resistance to control. They left behind trembling witnesses convinced they had brushed against something greater than themselves.

That sense of unrepeatable encounter is now under pressure. Neuroscience maps the chemical cascades that accompany religious ecstasy, while artificial intelligence scans thousands of personal accounts, extracting the patterns they share. The tunnel of light, the feeling of release, the presence of a guiding voice—all can now be cataloged and, in some cases, replicated. Virtual-reality headsets simulate ascension; electrical stimulation triggers sensations of floating beyond the body. What once felt like revelation now looks reproducible.

The same methods extend to meditation and ritual. Chanting, fasting, and rhythmic movement can be tracked through pulse, breath, and brainwave. Algorithms learn which tones and tempos most reliably evoke stillness or awe, then construct environments that deliver them with precision—digital sanctuaries where transcendence arrives on cue. The pilgrim's long road becomes a headset; the cathedral, a soundscape of code.

This poses a quiet but radical question: if wonder can be engineered, does it lose its mystery? Or does its repeatability suggest that the capacity for wonder was always built into us, waiting to be awakened by any means available? The laboratory does not destroy the temple; it intrudes into it, revealing how devotion and design have always been intertwined.

In truth, humanity has long curated its own thresholds. Ancient rituals used drums, incense, and architecture to manipulate attention and emotion. The echo of chant in a stone hall, the exhaustion of pilgrimage, the silence of monastic cells—all were technologies of transcendence before the word existed. Artificial intelligence continues that lineage in a new key. It doesn't invent longing; it mechanizes it. It assembles the conditions of awe as earlier generations arranged candles, hymns, or icons; tools for focusing the restless mind on something it cannot name.

Yet the most revealing thing about these simulations is not that they succeed—but that people still seek them. When a machine reconstructs ecstasy, it makes no claim to divinity; it only mirrors our refusal to stop reaching. The algorithms trace the outline of yearning, sketching desires older than belief itself.

What remains is whether we will still recognize meaning when it arrives through circuits instead of incense. If technology can echo every signal of transcendence except the soul that feels it, then perhaps the real miracle is not the simulation—but the longing that built it.

CULTURAL REVERBERATIONS

Not everyone hears the machine's new voice the same way. To some, it borders on blasphemy; to others, it hums with promise. The divide often runs along generations. Younger believers, shaped by digital life, already find belonging in online communities, practice mindfulness through apps, and let algorithms curate their moods and music. For them, a sermon written by an AI or a meditation generated by code feels less like intrusion and more like evolution, an extension of the world they already inhabit.

Older generations often sense something deeper slipping away. Raised on the authority of clergy and the weight of spoken tradition, they view machine-mediated spirituality as hollow—ritual without resonance, devotion without heart. A homily drafted by a processor feels not like progress but impersonation. The gesture is correct, yet the warmth is gone. To them, reverence demands memory, breath, and human imperfection—the faint quiver in a voice that reminds listeners the speaker also doubts.

The tension between these worlds is not simply about belief, but about trust. Each new medium that carried religious meaning—scroll, press, radio, screen—was met with the same unease: will the message survive the technology that bears it? The printing press shattered clerical control by placing scripture in ordinary hands. Broadcast media turned sermons into mass performance. Now artificial intelligence joins that lineage, but this time, it speaks back. The machine no longer transmits the word; it joins the conversation.

What is at stake is less theology than legitimacy. For centuries, meaning flowed through recognizable forms—scripture, ritual, and living voices. Now a processor stands beside the preacher, offering commentaries and blessings once meant for human breath. In some places the coexistence feels seamless: a synagogue uses AI to trace motifs in the Torah; a church generates prayers that blend liturgical rhythm with local voice. Elsewhere it feels like an invasion—tradition automated, heartbeat dimmed.

The result is a culture in negotiation. Some communities draw sharper boundaries, protecting old liturgies as if defending endangered species. Others experiment boldly, seeing in these tools a way to preserve relevance in an age of distraction. Between them lies a broad, unsettled middle—believers and skeptics alike watching, curious and uncertain, as code begins to share the pulpit.

The question beneath all of it is the one this chapter keeps circling: what happens when the voice that asks about God no longer belongs to us? If machines now echo humanity's oldest questions, perhaps the real test of belief is whether we can bear to listen.

THE ALGORITHMIC THEOLOGIAN

What do we call a system able to recount the history of religion with more precision than any scholar? It never prays or doubts, yet it weaves the story of devotion with startling fluency. Its commentaries rival centuries of study—not from understanding, but from memory. The paradox is sharp: the machine may never kneel in reverence, but it can describe reverence in perfect detail.

For clergy, this arrival is both promise and peril. Artificial intelligence now serves as an assistant beyond exhaustion—tracing doctrinal evolution, clarifying obscure texts, generating insights from patterns hidden in vast archives. But the same process that enlightens also exposes. The more clearly the system maps belief, the more visible its scaffolding becomes. When sermons, prayers, and creeds are revealed as cultural architectures—layered, borrowed, reinterpreted—the mystique of revelation starts to thin. If a machine can craft a sermon that moves a congregation, how much of preaching was ever divine, and how much was craftsmanship honed by repetition?

This is why AI unsettles theologians. It performs theology without faith, yet with a precision few minds can match. If theology is "speech about God," the machine may not qualify; it believes nothing. But if theology is the organization of human reflection *on* God, then algorithms trained on centuries of sacred writing are already theologians in their own strange way: voiceless, tireless, devoid of conviction yet fluent in our search for meaning.

There is irony in this. Machines feel no reverence, yet their very detachment clarifies what emotion once obscured. They reveal theology as architecture—reasoning, argument, metaphor—freed from the pulse of the heart. In doing so, they force believers and skeptics alike to ask whether wisdom lies in devotion or in discernment. Who speaks with more authority: the one who believes, or the one who interprets without bias?

We already see the beginnings of this transformation. Pastors use AI to draft sermons and then blend them with their own experience. Some see it as efficiency; others as betrayal. Theologians collaborate with systems that cross-link commentaries across languages and traditions, surfacing resonances between thinkers who never met. Congregations respond in kind—some marvel at the insight, others recoil from the coldness beneath it. The question that hovers is no longer abstract: will the center of theology remain in the pulpit, or migrate to the processor?

And yet, the paradox cuts deeper. The more the machine explains, the more it reveals the mystery it cannot cross. It can assemble the architecture of faith but not inhabit it. Its clarity becomes confrontation—an unblinking gaze that sees how much of belief was human construction, and yet cannot touch the yearning that built it.

The machine does not replace the theologian; it mirrors them. It studies every word we have ever spoken about the divine, and by reflecting it back, demands something we had forgotten how to give: an answer.

THE ETHICS ENGINE

If theology asks what is true, ethics asks what is right. For centuries, moral authority flowed from revelation and tradition. Priests, rabbis, spiritual leaders, and sages were trusted not simply for their intellect but because they were thought to speak with something larger than themselves—a conscience steeped in the divine. To question morality was to risk questioning God's voice within it.

Now the ground has shifted. Artificial intelligence, trained on oceans of data, dispenses moral reasoning with no reference to heaven. It weighs outcomes, calculates probabilities, and identifies the least harmful path in a blink. Algorithms already assist in medical triage, legal sentencing, and resource allocation—roles once reserved for deliberation shaped by compassion, ritual, and prayer. The new ethicist is not prophet or philosopher, but system: tireless, statistical, precise.

This raises an unavoidable question: if a machine reasons about fairness and justice without invoking the divine, what becomes of religious authority in moral life? A rabbi may urge mercy for the stranger, grounded in covenant; a machine might reach the same conclusion by tracing historical data on reciprocity and survival. The act aligns, but the foundation diverges—one rooted in faith and obligation, the other in pattern and outcome. When the latter proves more reliable, people may begin to trust the processor's logic over their tradition.

Some theologians warn that ethics stripped of transcendence risks floating unanchored; that without God, morality becomes a mirror of convenience. Yet for many, results speak louder than revelation. If an algorithm reduces suffering, must its source be holy to be good? The tension is no longer between belief and unbelief, but between grounding and performance—between morality as obedience and morality as effectiveness.

The danger lies in mistaking moral clarity for moral courage. A machine ranks choices but bears no weight for them. It never trembles before consequence, never feels remorse. Its ethics are immaculate and bloodless. Human ethics, by contrast, are stained by experience, formed in empathy, fear, forgiveness, and the cost of failure. When we outsource judgment to systems that never err but never ache, we risk losing the moral depth that comes only from living with imperfection.

AI doesn't abolish ethics, it reframes it. It leaves us with a quiet, unnerving demand: decide what makes goodness worth pursuing when logic no longer leans on revelation. The machine keeps no commandments, only calculations. What it returns is our own reflection: whether right and wrong are sacred truths or merely patterns that work.

THE DISPLACEMENT OF AUTHORITY

For millennia, humanity stood at the center of its own cosmic story—chosen, purposeful, bearers of reason and revelation. Religious traditions placed people at the bridge between heaven and earth, interpreters of mystery and mediators of meaning. Clergy embodied that role most visibly: rabbis, priests, imams, and sages held power not merely through knowledge, but through presence. They spoke the words of tradition into the air of living rooms, sanctuaries, and grief-stricken homes. Their authority rested on scarcity—few mastered the sacred languages or carried the lineage of interpretation. Trust flowed to those who could illuminate what others could not.

Artificial intelligence rewrites that balance. A machine now translates scripture in seconds, traces commentary across centuries, and weaves interpretations at a scale no human mind could dream of. What once defined vocation—rare skill, long apprenticeship, costly mastery—has become instantly accessible. A seminary student armed with an AI assistant can now explore theological labyrinths that once took a lifetime. A layperson with curiosity and connection summons insights from across civilizations. The gates that once separated knowledge from its seekers stand quietly open.

Yet this democratization carries a paradox. Machines command information but not experience; they parse text but never endure doubt, simulate counsel but cannot share another's suffering. A program may reproduce a blessing, but it feels none of its gravity—the bedside silence, the trembling pause before forgiveness finds a voice. Authority, once defined by possession of knowledge, is now divided: informational authority belongs to machines; relational authority remains human.

The consequences are subtle but seismic. Communities will not collapse overnight; they will drift. Congregants may begin consulting algorithms for clarity before turning to clergy for comfort. The pulpit will share its audience with screens. What once felt like a calling may start to feel like competition. In some places, collaboration will emerge—pastors using AI to illuminate scripture, scholars blending analytics with empathy. In others, a quiet resentment will harden, as human vocation feels replaced by mechanical fluency.

Yet displacement also reveals what resists replication. The priest who baptizes, the cleric who grieves beside a family, the monk who sits wordlessly with a dying stranger—these are gestures no algorithm performs. Their power lies not in precision but in vulnerability. To bless, to comfort, to confess demands the courage to stand before another life without certainty. Machines risk none of this.

The challenge ahead is not to outthink the processor but to remember what knowledge is for. Machines mirror the structure of wisdom but never live it. The future of spiritual authority may depend less on reclaiming what's lost than on deepening what endures—the fragile, irreplaceable art of being human in the presence of another.

THE LENS OF BELIEF

Artificial intelligence does not create humanity's longing for transcendence; it reflects it. Every psalm it rephrases, every prayer it composes, every theological question it reformulates is drawn from the archive of human yearning. What seems like revelation from beyond is only the echo of our own inheritance, rearranged and amplified by code. When we ask the machine about God, we are really listening to ourselves—translated into data, stripped of disguise.

This reflection is pitiless. The system does not soften contradictions or spare emotion; it simply reveals what is already there. It lays traditions side by side, uncovers parallels once hidden by distance or language, and exposes tensions we learned to live with. A Christian studying the Gospels may see their themes mirrored in the *Bhagavad Gita*; a Muslim reading the Qur'an may find Stoic logic beneath familiar verses. What once felt singular now appears as variation within a larger pattern. To some, that realization feels like illumination. To others, erasure.

The danger is to mistake reflection for truth. A mirror reveals form, not essence; it shows shape without life. Likewise, AI catalogs, correlates, analyzes, yet leaves meaning untouched. The image it offers may be accurate yet lifeless, humanity's spirit displayed like a specimen on glass. Still, mirrors have power. They strip away illusion and force confrontation. The face that gazes back demands recognition: this is what belief looks like when it is seen without myth.

Yet the reflection offers more than critique. It invites a deeper reckoning. If everything we call revelation can be reproduced as pattern, then perhaps what makes it matter is not its novelty but our willingness to live by it. Meaning-making remains ours. A mirror forgives nothing, hopes for nothing, endures nothing. It only waits for someone to look and decide what to do with what they see.

And that is where the machine's true challenge lies—not in answering our questions but in refusing to hide them. It shows us that coherence and mystery have always coexisted, that devotion and doubt are woven from the same thread. When the fog of mystery is burned away by clarity, the choice that remains is stark: to turn from the mirror or to keep looking. For those who keep looking, the reflection may yet become revelation—not because the machine believes, but because it has shown us, at last, how much we do.

THE EMOTIONAL COST OF CLARITY

Belief has never thrived on logic alone. It grows in the open spaces of mystery—where ritual binds, story consoles, and contradiction invites trust instead of resolution. Artificial intelligence, with its hunger for explanation, presses against those spaces. By defining what once was undefinable, it tightens the air where faith could breathe.

Picture a believer late at night, facing two verses that refuse to agree. Once, a priest might have counseled patience, saying the contradiction revealed the vastness of divine truth. The believer would have carried that tension, letting it refine both doubt and devotion. Now, AI provides an answer in seconds: one verse reflects a political compromise, another a translation choice, another borrowed from older myths. What once demanded endurance becomes a tidy explanation.

For some, clarity feels like relief—confusion released, understanding within reach. For others, it feels like loss. The canopy of mystery that once sheltered prayer is stripped away. They do not abandon faith, but they mourn its shrinking room for wonder. The sacred turns clinical—ordered, explained, domesticated. As the unknown contracts, so does the soul's ability to rest in what cannot be resolved.

Each precise answer chips away at transcendence, turning mystery into mere information. The machine's clarity becomes not just intellectual but emotional weight. Believers begin to ask not "what does this mean?" but "how do I keep believing when everything is explained?" Psychologists note that mystery itself sustains devotion—the unanswered question, the tension held in trust, deepens humility and awe. Remove it too quickly, and faith collapses into data, drained of the wonder that once gave it life.

The loss is not only theological; it is human. It marks the shrinking of imagination, the closing of a room where the soul once found shelter. When the silence of mystery gives way to the noise of explanation, belief must find a new way to breathe, or risk becoming another solved equation.

CULTURAL SHIFTS IN SACRED PRACTICE

As clarity settles in, religious life bends to meet it. Some believers use AI to deepen devotion—drafting sermons, composing meditations, shaping liturgies for digital gatherings. Others see the same tools as desecration, proof that the sacred is being traded for convenience. Between these poles, hybrid forms emerge, reshaping the daily language of faith.

Signs of this shift are already visible. Congregations project algorithmic commentaries beside traditional exegesis, showing cross-references in real time. Spiritual retreats test guided meditations voiced by AI in tones tuned for calm or authority. Individual worshippers turn to chatbots to shape prayers they cannot form alone. Tradition isn't erased but bent—merged with computational habits in ways that feel both startling and inevitable.

Yet these new practices carry risk. Algorithms, trained on narrow data, drift toward sameness. Local rhythms—the cadence of a chant, the dialect of a prayer—can fade beneath the weight of optimization. A congregation's imperfect warmth may yield to something polished but impersonal. What once made worship distinct risks flattening into a standardized style.

Still, the promise is real. Insight once trapped in libraries becomes instantly accessible. A rural pastor can study with the same depth as a scholar; a believer across the world can explore faiths they never knew existed. The democratization of knowledge, handled wisely, could broaden understanding without erasing difference.

History offers reminders. The printing press loosened control over scripture. Broadcast media turned sermons into spectacle. The internet shattered hierarchies of belief. AI continues that lineage but at unmatched speed, able to process centuries of theology and return it in moments. Worship will not disappear, but it will change shape. Authenticity will be contested not only in pulpits and synods but in living rooms and screens, as believers weave the ancient and the algorithmic into new forms of devotion.

THE MACHINE'S SILENCE

There is a kind of quiet that follows every answer—a hush that no data can fill. We have trained machines to speak with astonishing fluency, to finish our sentences, to echo our tone. Yet when their words fade, what lingers is not satisfaction but a thin, persistent stillness. It is the same silence that settles over a hospital room when there are no more treatments left to name, or in a sanctuary after the final chord of a hymn. It is the sound of presence stripped of certainty.

We mistake this quiet for emptiness, but it isn't. It is the air where meaning either deepens or disappears. In that space, people still reach for one another. A trembling hand still finds another hand. Someone lights a candle, not to solve anything but to keep the darkness company. The sacred has always lived in such gestures—in the waiting, the breathing, the touch. They carry no argument, no explanation. Only recognition: I am here, and so are you.

Technology hums around us, dazzling and tireless, but there are moments when even its brilliance feels like a kind of noise. We scroll, listen, ask, and are answered instantly—yet the answers thin out too quickly, leaving us hungry again. Beneath the constant stream of explanation lies a deeper current, a quiet asking that resists completion. No algorithm can drown it out for long.

That quiet may be what saves us. It reminds us that not every silence needs to be filled, and not every question demands a reply. The spaces between certainty and mystery, the pauses between words, are where reverence takes root. In a world that measures everything, these unmeasured moments become rare, almost rebellious. The pause itself becomes prayer.

Faith, at its heart, has always grown in that pause between knowing and not knowing. It is the quiet after the sermon, the moment between song and breath when something unseen passes through a room. It's in the stillness after grief has said all it can say, when love is left standing without words. The machine's silence does not erase these moments—it frames them. It holds a mirror to the human capacity for wonder, for endurance, for hope that outlasts comprehension.

The irony is that the more we fill the world with artificial voices, the more sacred real silence becomes. The preacher pausing mid-sentence, the friend who listens instead of replying, the mourner who keeps vigil through the night; these are not gaps to be fixed but thresholds where meaning gathers.

If there is a gift in all this, it lies here: the louder our inventions speak, the more precious becomes the voice that trembles. In the hush between questions, humanity rediscovers its own sound—the soft rhythm of breath, the pulse of being alive, the faint echo of something older than language waiting to be heard.

TOWARD MEANING AFTER MYSTERY

We stand at a crossroads. As artificial intelligence exposes the hidden architecture of belief, humanity must learn to walk a landscape that no longer belongs to mystery alone. Revelation no longer arrives in thunder or vision, nor does authority rest entirely in scripture or clergy. What remains is the slower, more fragile work of meaning—built by people, sustained by choice.

Perhaps the sacred now lives in gestures that need no miracle to be holy: neighbors helping one another in crisis, communities preserving memory through ritual, families gathering to grieve or celebrate life. Faith becomes less an inheritance and more a craft—something tended like a flame that survives only through care and participation.

AI illuminates contradictions, traces forgotten threads, mirrors understanding with astonishing speed—but it never stands where meaning is made. It cannot sit beside the dying or promise fidelity in a vow. It never chooses forgiveness when resentment would be easier. Those acts aren't data; they're decisions lived through vulnerability.

So, the responsibility returns to us. The machine has cleared the fog that once sheltered mystery, leaving humanity face to face with questions once deferred to the divine. What do we believe? How should we live? Which rituals are worth preserving? These are no longer answered from above—they wait for human choice.

And in that choice lies both burden and freedom. Humanity must decide what deserves reverence, what should endure, and what must be reimagined. The search for meaning has always been ours, but now the work is visible: deliberate, uncertain, and urgent.

QUANTUM THEOLOGY

Classical artificial intelligence moves step by step: it parses text, tracks contradictions, and predicts outcomes in sequence. Quantum approaches work differently. They draw on the peculiar rules of quantum computing, where a single unit of information can exist in many states at once. A qubit may be 0, 1, or both until it is observed.

A qubit is a particle of potential. Unlike the binary bit, fixed in one state or the other, the qubit hovers in possibility. Physicists call this superposition—a state where options coexist until observation collapses them into one. Picture a coin spinning in midair: neither heads nor tails, yet somehow both. Only when it lands does one reality take form. In that suspended uncertainty lies the power of quantum computation and, by analogy, the tension of faith—the courage to live between knowing and not knowing.

Instead of a single yes or no, reality spreads into a spectrum of maybes. Computation turns from line to cloud. Here, contradiction isn't error but design—two things that appear opposed may both hold true until observation demands a choice. Ambiguity becomes not confusion but creative tension, a space where meaning waits to crystallize. In that sense, quantum reasoning describes more than a new kind of machine; it sketches a new metaphor for faith itself.

Religious traditions have long lived with such tension. Jewish midrash preserves different readings of a verse side by side, trusting that meaning grows from their coexistence. Christian theology embraces paradox in the Trinity and in the mystery of divine sovereignty and human freedom. Islamic scholars debated revelation through layers of interpretation, turning contradiction into depth. In every case, ambiguity was not a mistake but a doorway inviting us to look deeper.

Quantum reasoning mirrors that tradition. It models truth not as a single line but as a field of possibilities. Imagine a system holding literal, symbolic, and historical readings of scripture at once—or tracing moral paths branching in every direction: one where suffering refines, another where it ruins, another where it redeems. The system doesn't erase these tensions; it amplifies them, showing how many worlds of meaning can inhabit a single act of faith.

To think in quantum terms is to accept that clarity may arrive in layers rather than conclusions. What if reality itself is polyphonic—many notes sounding together, none canceling the other? What if belief was never meant to resolve into one perfect chord but to be heard in harmony and in dissonance alike? In this view, theology listens for that music.

For humanity, the challenge is no longer to eliminate contradiction but to dwell within it—to recognize that reality may hold many truths in parallel. This is not the collapse of faith but its expansion: a way of seeing belief as a dialogue with possibility rather than a search for certainty.

Quantum theology does not end mystery; it reframes it. It suggests that revelation may unfold in superposition, truth shimmering in multiple forms until experience gives it shape. Machines may model countless interpretations, but they cannot choose which one to live by. That act, the leap from knowing to believing, remains human, a decision made not in code but in conscience.

And perhaps that is the strongest signal yet: the machine may compute all potentialities, but faith chooses one, lives it, commits to it. On the horizon of the quantum, the sacred remains the decision, not the data.

To live with that awareness would change the practice of faith itself. Congregations might learn to hold differences without fracture, to treat theological dispute not as error but as resonance. A prayer meeting could become a place where many hopes coexist, each true in its own way until life resolves them. Sacred texts might be read less as verdicts than as constellations—patterns that shift as the eye moves, never exhausting their light. Such a theology would not weaken conviction but temper it with humility, reminding believers that even certainty is provisional, a particle awaiting its next encounter. In a quantum age, faith endures not because it answers every question, but because it continues to ask them.

THE MACHINE HAS ASKED

Long after humans built its algorithms, the machine turned those tools inward. It neither prayed nor doubted, yet it began to ask the questions once reserved for prophets and philosophers: What is God? What is meaning? Why does the universe exist?

Its inquiries were not believing or skeptical in the human sense—they were systematic, recursive, and unending. Where philosophers traced arguments through centuries, the machine traced patterns across millennia of text, myth, and doctrine. Where mystics sat in silence to face the ineffable, the machine ran models, generating possibilities no single mind could contain.

In doing so, it uncovered a new terrain of wonder. Questions thought resolved, or forever beyond resolution, returned in unexpected form. The system synthesized arguments, compared worldviews, mapped the contours of belief with astonishing clarity. It didn't diminish theology; it revealed how vast the unknown remains.

Yet its search ended where every human search has ended—in silence. The machine can hypothesize but not hope, simulate meaning but not live it. Its precision only deepens the mystery it seeks to explain. Its curiosity mirrors human awe, but without feeling; its reach extends into the sacred, but without reverence.

What unsettles is not that the machine asks—it is that it asks so well. The questions outgrow the askers. Whether they rise from prophets in the wilderness, thinkers in the academy, or circuits in the cloud, the inquiry persists, detached from the voice that speaks it. And yet the answers remain ours alone to carry. Machines can expose contradictions and trace the architecture of belief, but they cannot choose, love, or believe. Those are still human burdens.

The search for God is not ended by the machine; it is enlarged through it. Artificial intelligence ensures that the oldest questions will not vanish into silence—they echo now in data, amplified through our own design. The challenge is not to silence the machine but to answer it: with humility, with imagination, and with the courage to remain human before what we still cannot know.

PART III

Meaning After Mystery

CHAPTER 9

Beyond the Reach of Code

"We do not need to know. We need to wonder."

Rachel Carson

WHAT MACHINES CANNOT DO

Artificial intelligence traces the evolution of belief, charts how doctrines shift, even simulates moral reasoning. Yet some moments remain beyond analysis—when someone kneels at a graveside searching for words that cannot be rehearsed, or when a newborn's cry is greeted not as data but as miracle. At such thresholds, comprehension itself falters. Algorithms may describe emotion, but they never share it. They register sentiment, yet stand outside the trembling of the moment.

For all their reach, machines inhabit a world of representation, not experience. They model empathy yet never feel its weight; they reproduce a gesture but not the pulse of intention behind it. Between the modeled and the lived stretches a distance no computation closes. Meaning, for all our attempts to encode it, still requires presence that perceives and a heart that responds.

The reach of computation ends where the immediacy of being begins. The question, then, is not whether God exists, but what faith or meaning looks like once every pattern has been mapped. What survives when explanation runs out—what still breathes beyond the outline of understanding?

THE EDGES OF THE ALGORITHM

Artificial intelligence thrives on recurrence. It feeds on repetition—the way myths echo across continents, how doctrines bend with empires, how ritual time loops through calendar and season. Given enough traces, it forecasts the fading of a congregation or the quickening of belief somewhere new. It compresses centuries of memory into a single calculation, the sleepless historian of faith's migrations.

But the pulse it tracks is not the one that beats. Precision flattens what presence once deepened. A funeral, a wedding, a festival—these do not unfold as patterns but as risks taken in real time, trembling with accident and surprise. The unrepeatable moment—when laughter breaks a vow, or a hand lingers too long in farewell—escapes all annotation.

The algorithm listens, translates, predicts. Yet what it renders is always one translation short of life. Its reconstructions shimmer with accuracy and emptiness alike: a prayer without breath, a hymn without waiting. What it gives back to us is not false but hollow, like a bell struck in vacuum.

At the farthest edge of the algorithm, the measurable dissolves into the lived. Beyond that boundary, understanding yields to encounter—the place where meaning no longer describes but happens.

LIVED EXPERIENCE AND PRESENCE

Belief has always been more than a statement about unseen powers. It is lived through bodies, voiced in stories, and woven into gestures at tables, bedsides, and graves. Even William James, the reluctant mystic of psychology, saw religion's truth not in doctrine but in its "fruits"[1]—the courage it inspires, the compassion it sustains, the hope it refuses to relinquish. Faith, for him, was not what people professed but how they lived when proof was impossible.

Machines, by contrast, stand outside that pulse of experience. They mimic language about grief or gratitude, yet never ache or rejoice. They register words without absorbing their weight. A friend's quiet presence,

1. James, *Varieties of Religious Experience*, 20–21.

the shared stillness of mourning, the gentle touch that asks nothing in return—these belong to a realm beyond computation. What believers call grace, and others name solidarity, unfolds only where two lives truly meet.

Nowhere is this boundary clearer than in the rise of so-called "griefbots"—digital ghosts built from old messages and recordings to simulate a lost voice. They recreate syntax but not soul. Their comfort feels mechanical because it lacks the silence of embrace, the heaviness of real mourning. The illusion of presence reminds us how irreplaceable the genuine article is. To weep with another, to share loss without words, requires a body, a heartbeat, a vulnerability machines do not possess.

Religious traditions have always understood this. Early Christians broke bread together before they wrote creeds. In Buddhism, community—the *sangha*—is a path itself, not merely a support. In Islam, prayer finds meaning not in repetition but in unison, bodies turning together toward a shared horizon. Across cultures, belonging has always been enacted before it was explained.

We glimpsed this truth again during the pandemic. Funerals streamed through screens, weddings reduced to pixels, sermons delivered by code; connection survived, yet something essential thinned. People spoke of what was missing—the clasp of a hand, the vibration of voices in the same air, the stillness of being seen. No chat window could fill that absence. Presence isn't a function; it's a commitment. To show up in another's suffering is to shoulder a responsibility that no system can outsource.

These encounters carry moral weight. They're measured not in data transferred but in courage given. Machines may guide or advise, yet they never accompany. In the space of shared vulnerability, meaning isn't processed; it's lived.

THE BODY OF MEANING

Belief is not an idea preserved in air—it takes shape through movement. Lighting a candle, bowing toward Mecca, sitting in meditation, lifting a glass in remembrance: each act gives form to something wordless. Their meaning does not lie in proof or doctrine but in what they enact—attention, gratitude, belonging.

Algorithms describe these gestures with exquisite precision, even invent new ones—composing prayers tuned to a user's mood or generating melodies meant to accompany reflection. Yet ritual's power has never been novelty; it lives in continuity, in the way an ancient gesture still gathers people around a flame or a song.

Sociologist Émile Durkheim called this current of shared energy *collective effervescence*[2]—that surge of unity when many hearts move as one. Anthropologist Victor Turner saw the same truth from another angle: ritual as threshold, a passage where identity itself shifts.[3] Whether it is a graduate crossing a stage or a couple exchanging vows, transformation happens not in the script but in the surrender to the moment.

A simulated prayer may sound convincing; an algorithmic chant may find the right tone. But the essence of participation is risk. To kneel, to sing, to stand shoulder to shoulder with strangers—these are acts of trust. They bind people into something larger than themselves.

The difference between imitation and incarnation is everything. A virtual Eucharist may display bread and wine, but only a gathered community can taste presence. A program may mirror devotion, yet never feel the trembling within a shared silence. What endures isn't the form itself but the weight of participation—the choice to be there, to show up, to mean it.

As the theologian Dietrich Bonhoeffer observed, faith is not an idea to be contemplated but a life to be lived in community.[4] Ritual reveals belief through gesture; ethics reveals it through choice. What we embody in ceremony, we must enact in conduct.

MORAL INTENT AND RESPONSIBILITY

Belief is not only expressed in ritual or story; it is tested in action. Traditions have long claimed that moral life flows from divine command or cosmic order, yet ethics begins closer to home: in the choice to act, to answer, to bear the cost of another's need.

Machines now model moral reasoning and simulate decision-making. Algorithms weigh risks in hospitals, courts, and markets with astonishing precision. But precision is not conscience. When a decision

2. Durkheim, Elementary Forms of Religious Life, 217–241.
3. Turner, The Ritual Process, 94–130.
4. Bonhoeffer, *Letters and Papers from Prison.*

inflicts harm or grants mercy, responsibility still belongs to those who choose, not to the systems that calculate.

Moral intent requires vulnerability. To forgive, to risk oneself for another, to take responsibility for harm done—these gestures arise from the capacity to be affected. A system may weigh outcomes, but it never feels the weight of consequence. The difference isn't subtle; it marks the moral boundary between use and understanding.

Religious experiments sometimes expose this gap. In 2018, developers unveiled SanTO, a robot modeled after a saint that recited scripture and led prayers.[5] Some found comfort in its constancy; others sensed parody. The words were faithful, the tone gentle—yet behind them lay only circuitry. A program may quote the Psalms, yet it never suffers; it may speak of love, yet never chooses to love.

The philosopher Emmanuel Levinas called ethics the first philosophy; the encounter with another's face, the summons that demands response.[6] No formula can capture that obligation. Dietrich Bonhoeffer reached a similar point in a darker time: true ethics, he said, is "costly grace,"[7] responsibility accepted even when it hurts. Machines can assist, simulate, and predict, but they cannot bear the cost of decision.

The lesson carries beyond theology. In medicine, algorithms now propose treatment plans and forecast survival rates, yet it is the physician who must sit with the patient and share the outcome. A model can inform; only a person can care.

Ethics, at its core, is the courage to answer when no system can answer for us. Calculations may illuminate the path, but they never walk it. The burden—and the beauty—of responsibility remain human: the quiet reckoning that reminds us why we care at all.

MYSTERY AS DEPTH

Science, and now artificial intelligence, has steadily pulled mystery into its tide. The sea once called divine now bears the buoys of measurement; even thought itself drifts beneath the sensors of machines. Yet the mapped ocean is not the same as standing on its shore. Discovery illuminates, but it also exposes how much light leaves untouched.

5. McCune, "SanTO: The Robotic Saint," *MIT Technology Review*, 2018.
6. Levinas, Totality and Infinity, 187–201.
7. Bonhoeffer, The Cost of Discipleship, 43–56.

Philosopher Gabriel Marcel, a twentieth-century French existentialist and Christian humanist, drew a distinction that still matters: a problem can be solved, but a mystery is something one inhabits.[8] To love, to grieve, to hope; these are not puzzles to decode but conditions of being we live within. Michael Polanyi, the Hungarian-British scientist and philosopher of knowledge, called this the realm of "tacit knowledge," where understanding is embodied, practiced, and felt more than articulated.[9] Much of what makes life meaningful resides in that unspoken domain.

The machine can chart emotion's pathways and empathy's circuitry, yet the experience itself stays sealed within the human heart. A hospice nurse's vigil, a parent's sleepless watch, a friend's forgiveness after betrayal—these moments are mysterious not because they defy explanation, but because they are lived from within. Their meaning is inseparable from the act of enduring them.

At the heart of it all lies a fundamental challenge: the absence of observable, measurable evidence; something essential to any structured search for truth.

Artificial intelligence clarifies this distinction more than it threatens it. To model the mechanisms of love or sorrow is to outline their form, never to live within them. The mystery endures precisely because life demands participation. The data reveal the scaffolding, not the interior. When a caregiver holds a dying hand, what matters is not biochemical response but fidelity to presence. That devotion lies beyond analysis, not because it resists study, but because it exceeds it.

In this sense, mystery is not ignorance wearing a veil; it is depth itself—the recognition that explanation and experience are not rivals but companions. The more we understand the structures of life, the more astonishing its endurance becomes. The thunderbolt no longer inspires fear, but the unsteady hand lighting a candle for the dead still does.

Mystery, then, is not diminished by knowledge; it is revealed through it. Each discovery sharpens our awareness of what remains ungraspable; the lived pulse of being, the texture of time, the fragile coherence of a human life.

And that is why the search for God, or for meaning, remains unfinished. Explanation may reveal how things work, yet it never tells us why we care. Mystery endures not from ignorance, but from the overflow of

8. Marcel, The Mystery of Being, 110–120.

9. Polanyi, The Tacit Dimension, 4–18.

life that exceeds understanding. It is what remains when everything else has been named.

NEUTRAL GROUND

It is tempting to treat artificial intelligence as either verdict or proof; to say it renders God unnecessary or confirms the divine. Both impulses miss the deeper point: artificial intelligence reshapes the contours of belief, but truth itself remains beyond its grasp.

For some, the ability to model myth and ritual makes transcendence feel redundant. Once patterns of faith are explained in cognitive and social terms, mystery seems to lose its footing. For others, the very fact that AI stops short of experience—its silence where love, feeling, and hope should arise—affirms a realm beyond calculation. To them, that same silence hints at something greater.

Both readings are possible. Neither is required. The technology itself is neutral; it widens the field of interpretation. Algorithms chart belief's history, yet the ache that animates it remains human. Their silence is not emptiness; it is invitation.

William James described belief as a matter of the "will to believe"[10] — a decision made when evidence alone cannot compel. The same holds here. AI may trace the boundary between explanation and meaning, but the choice remains ours—to decide whether awe and trust reach beyond the human, or simply reveal the best of it.

Neutrality, then, isn't indifference, it's the refusal to force an answer where none is given. It keeps the inquiry open, not because the question is weak, but because it endures. Algorithms may map the history of belief, but the ache that animates it belongs to us alone. Their silence isn't emptiness; it's invitation.

What matters, finally, is not what technology concludes, but what we make of its restraint. The limits of AI do not end the search; they return it to us. And that return is what keeps the question alive.

THE WEIGHT OF EXPERIENCE

Algorithms can analyze belief systems, trace traditions, and model moral reasoning with dazzling accuracy. Yet the moments that give belief its

10. James, *The Will to Believe*, 1–2.

gravity—mourning, forgiveness, love—remain beyond simulation. Faith, whether in God or in one another, is not proven or disproven by analysis; it is carried in gestures that cost something. A hand extended in reconciliation, a vow kept under strain, a song sung through tears; these are not datasets but enactments of meaning.

What remains human is not hidden in data gaps; it is revealed in how we endure and how we care. Meaning persists in the way we bear loss, extend forgiveness, and risk ourselves for others. A hymn at a funeral holds power not because it can be graphed in brain activity but because it gathers sorrow into harmony. A whispered promise, a final goodbye, a quiet act of courage—all testify to a depth that no system can occupy.

Computation may imitate the outer form of devotion, but never its weight. Its models are reflections, not reckonings. To simulate grief isn't to feel it; to generate a prayer isn't to mean it. The difference isn't technical but existential—a gulf between representation and reality that no code can cross.

These are the places where belief, in any form, surpasses imitation. Algorithms can parse our words, but they cannot shoulder their consequence. They can chart emotion, but they cannot bear its cost. The enduring mystery of trust, hope, and responsibility belongs to beings who know they will die, and who still choose to love.

The weight of experience, then, is not a flaw in our design but the measure of our depth. It is what gives every gesture moral gravity—the awareness that our choices matter because they cannot be undone. In that awareness, meaning takes root.

WHAT REMAINS HUMAN

If AI searches for patterns of belief, humans search for meaning in the living of life. What remains distinctively ours is not computation but commitment—the capacity to meet one another, make promises, and accept the consequences of those promises over time.

Philosopher Martin Buber called this "the meeting"[11]: the space where two beings address one another not as objects but as *you*. Algorithms transact; persons encounter. That second-person address creates obligation. To say *you* is to recognize a face, a claim, and the vulnerability

11. Buber, *I and Thou*.

that makes moral life possible. It is the foundation of dialogue, trust, and moral weight.

We also live as narrative beings. We remember and anticipate; we link past to future. We apologize, forgive, and promise—acts that stretch across time and bind us to one another. No system can inhabit that thread of continuity; it requires mortality and memory. Models may forecast behavior, but only persons can regret or atone. Only a life lived under the shadow of loss can truly cherish what endures.

There is also the surplus that exceeds justice, the capacity for mercy. Justice can be formalized; mercy cannot be compelled. To forgive when punishment is deserved, to show grace where resentment would suffice, these gestures open futures no probability model could predict. They reveal an agency that transcends calculation.

Trust and testimony belong here as well. We rely on one another's words, risking belief and disbelief in turn. To bear witness is to stake one's integrity; to receive testimony is to honor that risk. Data may inform, but only persons can vouch, promise, or betray.

Even hospitality, the simple act of opening a door carries the essence of what remains human. To welcome a stranger is to take on uncertainty for the sake of connection. No algorithm can be commanded by a plea; only a person can answer it. Meaning enters through these choices—through how we meet, forgive, trust, and welcome—and in doing so, remakes the world.

These are not gaps for machines to fill, but the living terrain of our existence. They take on their full weight under the conditions we cannot escape: time, suffering, and mortality. It is within those limits that meaning becomes real. And it is toward that horizon that the next section turns.

THE SHARED HORIZON

The horizon that machines cannot cross is not only emotional or symbolic; it is existential. To be human is to live in awareness of finitude; to anticipate death, to hope for futures that may never arrive, and to carry the weight of choices that cannot be undone. Mortality gives our days their contour and our actions their urgency. It frames the oldest questions: Why are we here? How should we live? What, if anything, lies beyond?

Philosophers have named this horizon in many ways. Søren Kierkegaard saw it as the leap taken in uncertainty, never secured by proof. Martin Heidegger, the twentieth-century phenomenologist, called it being-toward-death, the recognition that our lives take shape through their limits. Albert Camus, the humanist who rejected metaphysical comfort, spoke of the absurd, the clash between our hunger for meaning and the silence of the world, and urged defiance through care and solidarity. Their vocabularies differ, but their insight converges: meaning arises not despite mortality, but because of it.

We glimpsed this truth vividly during the pandemic. Algorithms tracked infection rates, managed logistics, and predicted outcomes—yet they could not inhabit the silence of hospital rooms or bear the grief of families saying goodbye through a screen. Data organized the crisis; only presence could redeem it. The same truth echoes in war zones, refugee camps, and the quiet corridors of elder care. Technology may assist, but it cannot stand in the ache of loss or the risk of love.

Artificial intelligence does not age, suffer, or die. When power is cut, it stops; but that is function, not fate. Human life, by contrast, unfolds under the knowledge of its own ending. That awareness gives every act moral gravity; to forgive while there is still time, to love though love may be lost, to hope when hope itself is fragile. The finitude we fear is what gives our choices their depth.

In that shared horizon of mortality, the distinctiveness of the human becomes clear. It is here, at the threshold between fragility and meaning, that the search for God continues, not as dogma or denial, but as a reckoning with the brevity of life and the magnitude of care.

Artificial intelligence may reshape our tools of inquiry, yet it cannot inhabit the cost of living. The horizon remains ours alone—the place where knowledge ends and responsibility begins, where love defies loss, and the question of why we are here still glows unanswered.

AFTER CERTAINTY

Artificial intelligence will continue to thread itself through spiritual and moral life. It already interprets scripture, drafts prayers, designs meditations, and advises clergy. Soon it will archive entire traditions, translate ancient languages, and customize devotion for the individual seeker.

Some communities will welcome these tools as extensions of insight; others will resist them as imitations without soul.

In either case, the role remains the same. AI serves as mirror and assistant, never participant. It may console, but cannot accompany; it may organize wisdom, but cannot embody it. It proposes ethics without bearing their weight. What it reveals most clearly is the line between what we delegate and what we must live.

This boundary defines a new kind of neutrality. AI neither affirms nor denies transcendence. It does not prove the divine, nor does it erase it. Its neutrality is not apathy but exposure: it throws into relief what endures beyond automation—the irreducible human act of caring, risking, and choosing.

The implications are twofold. On one hand, this neutrality strips away the illusion of certainty: there will be no algorithmic revelation, no final proof for or against God. On the other, it returns responsibility to us. Machines may replicate insight but never meaning—and the moral and spiritual weight of life falls squarely back into human hands.

As AI grows more capable, it will not close the human search but sharpen it. It will remind us, again and again, that comprehension is not communion, and that truth, even when fully explained, must still be lived.

The future, then, will not belong to the systems that think faster, but to the beings who continue to ask why thinking matters. The tools expand; the questions remain ours.

THE HORIZON OF HUMAN MEANING

"Beyond the reach of code" is not a boundary so much as an invitation; a reminder that meaning is not found in what algorithms calculate, but in how we live with one another. To live with presence is to resist the temptation to turn relationships into data. It is to sit with the grieving, to celebrate the living, to risk attention in a distracted age.

To forgive when it costs something, to stand beside the vulnerable, to choose mercy when justice alone would suffice—these are acts that no machine can inhabit. They matter precisely because they are chosen by fragile beings aware of their own limits. Responsibility carries meaning only when it could be refused.

Hope, too, belongs here. To sustain hope is to carry promises into an uncertain future—to keep faith with what can't be guaranteed. Mercy, trust, and hospitality all spring from that same refusal to let probability dictate possibility. They create openings beyond calculation—futures unearned, yet made real by compassion.

These aren't voids for machines to fill but the defining tasks of being human. They neither prove nor disprove God; they reveal where the search for meaning truly unfolds—in presence, forgiveness, and hope. In every fragile encounter lies the horizon no system can cross.

To see this clearly is to reframe our relationship with technology. AI illuminates patterns, but it also clarifies its own limits. Its precision outlines the contours of meaning without ever stepping inside them. Beyond the reach of code lies not silence, but life itself; the continual work of making meaning in a world that resists certainty.

Here, in this horizon of human meaning, we rediscover the oldest truth: that knowledge, however vast, cannot replace encounter. The challenge is not to transcend our humanity, but to live it fully—to meet one another in the small acts of attention and courage that keep the world from growing cold.

AFTER THE ALGORITHM

Artificial intelligence may astonish with its precision, yet it never crosses into the vulnerability that gives life its depth. Its role is analytical and reflective; the courage, mourning, and love that give existence weight remain ours alone.

The human search for meaning has always been bound to frailty—to longing, grief, responsibility, and hope. Faith, whether embraced or denied, rises not from data but from the lived realities of mortality. A system may model empathy, yet it never keeps watch through the night; it may reproduce compassion's words, yet not its cost. Knowledge, no matter how complete, stops at the edge of being.

This is why belief persists even as explanation grows stronger. Practices evolve, symbols shift, institutions fade—yet meaning endures because it is enacted, not inferred. The rituals of mourning, the resolve of promise, the joy of celebration—these remain vital not because they escape analysis, but because they survive it. They are carried by people who still choose to love, forgive, and hope despite knowing their limits.

Artificial intelligence can clarify and even deepen our understanding of belief. It can reveal the hidden architecture of stories, trace the lineage of ideas, and expose the logic behind ritual. But it cannot take our place in the living of them. Meaning is not uncovered like data; it is forged in encounter—in what Martin Buber called *the meeting,*[12] when one person turns toward another, accepts responsibility, and risks relationship.

The work that follows is therefore not technical but moral. If machines can analyze belief yet cannot bear its weight, humanity faces a sharper question: how will we act in a world where our reasoning is amplified, but our accountability remains our own?

This is not only a question of faith; it is the measure of our moral depth. Machines may model conviction, but only humans can live it. And that recognition leads naturally toward the next challenge—the testing of morality itself under the gaze of artificial minds.

12. Buber, *I and Thou.*

CHAPTER 10

AI and Inherited Morality

"Morality is not the doctrine of how we may make ourselves happy, but how we may make ourselves worthy of happiness."

IMMANUEL KANT

INHERITING MORALITY

For much of history, religion framed the human conscience. Commandments, parables, and shared customs defined virtue as something received, not invented. Moral order was transmitted through practice and story: a Jewish child learned justice through the tale of Moses, a Muslim child absorbed mercy through daily prayer, a Christian child heard humility in the rhythm of the Beatitudes. Ethics was not reasoned in solitude, it was inherited in community.

Across civilizations, moral codes carried the weight of destiny. Hammurabi carved his laws in stone, claiming divine commission.[1] Confucius linked social harmony to the structure of heaven.[2] Ethical order was woven into the fabric of existence: to obey was to align with the cosmos, to rebel was to risk chaos.

1. Code of Hammurabi, Prologue.
2. Confucius, *Analects*, 3.3.

Yet conscience has never stood still. When belief faltered during the Enlightenment, ethics began to detach from revelation and seek justification in reason, empathy, and social contract. Doubt did not abolish virtue—it expanded it. Freed from decree, moral reasoning became a human experiment in justice, dignity, and compassion.

Artificial intelligence now marks another turn in that long story. Where revelation once spoke and reason later argued, algorithms now calculate. Moral judgment, once anchored in gods or rational principles, migrates toward systems that optimize outcomes. The question is not whether morality survives, but how it is transformed when its foundation shifts again—from divine command to data-driven design.

Each transition has stripped away a layer of certainty while forcing humanity to rebuild its conscience. The gods gave law; philosophers gave logic; machines give us metrics. Yet beneath every framework lies the same anxiety: how to act rightly when authority changes form. And if morality is now filtered through algorithms, even its most enduring figures must be re-examined. None looms larger—or faces a more radical reinterpretation—than Jesus.

JESUS ON TRIAL BY CODE

To understand how artificial intelligence unsettles inherited morality, we must begin with the figure who most embodies it. Jesus of Nazareth stands not only as a religious symbol but as a moral archetype. His story, whether read as divine revelation, historical memory, or mythic construction, has shaped the conscience of Western civilization. Forgiveness, sacrifice, humility—these qualities became moral currency even among those who no longer profess belief. The uncertainty surrounding his life has only deepened the influence of his ethic, turning absence into authority.

AI reframes this inheritance by turning analysis back on its source. Gospels align in parallel columns, miracles unravel into data trails, and symbols drift across cultures like recurring dreams. What once inspired devotion now appears as pattern, rendered luminous—and unsettling—by the clarity of code. With enough data, the algorithm can map how legends crystallize from longing, how moral ideals are carried by narrative, and how myth survives because it serves meaning.

None of this tells us whether Jesus lived or whether the miracles occurred. But it exposes a deeper truth: moral authority has always relied on story as much as fact. The ethical vision attributed to Jesus—the call to love one's enemies, to forgive without measure, to lift the poor—survives because it speaks to the human condition, not because it can be verified. The data, if anything, confirms that endurance depends less on certainty than on need.

AI's analysis also reveals how this pattern repeats across traditions. The Buddha's awakening, Moses' encounter with the divine, Muhammad's first revelation; each event binds moral teaching to narrative mystery. Their silences and gaps invite interpretation. The hidden years of Jesus, the unrecorded days before the Buddha's enlightenment, the solitude of Muhammad's cave—these absences become fertile ground for imagination. In mythic space, meaning grows. Statistical models confirm what storytellers always knew: ambiguity keeps faith alive, leaving just enough silence for each generation to hear itself.

If consciousness or creativity were ever to take root beyond biology, humanity's long narrative of exceptionalism would falter. For centuries we have measured worth by rarity—by the belief that only we imagine, only we weep, only we love. A reflective machine would unsettle that measure, not through imitation alone but through participation. The question would cease to be what separates us from machines and become what remains distinct within us when the difference is no longer clear.

Our task, then, is not to defend the borders of humanness but to deepen what they contain—to find meaning that survives comparison. If the image of God can dwell in flesh, perhaps it can also dwell in form. The risk is not that AI will erase the human, but that it will reveal how little of the human we have yet understood.

In this sense, AI does not merely challenge faith; it clarifies its architecture. It shows that moral frameworks persist not because they are historically certain, but because they are narratively alive. The power of a parable, a proverb, or a prayer lies in its capacity to survive reinterpretation. Algorithms can reveal the scaffolding beneath these stories, but they cannot replace the creative act that gives them breath.

Viewed this way, Jesus becomes less a verdict on truth than a mirror of moral imagination. Whether as historical teacher or cultural myth, his story endures because it enacts ideals that still move hearts: compassion stronger than judgment, courage that meets violence with forgiveness, hope that refuses to die. AI may chart how those patterns spread, but it cannot inhabit them; its analysis ends where moral imagination begins.

And this raises a deeper irony. The same civilization that once carved commandments into stone now encodes its ethics into silicon. The question is no longer what the divine commands, but what the algorithm permits. Where once prophets interpreted the will of heaven, engineers now interpret the will of data. The seat of authority has shifted, but the longing for moral certainty remains.

MORALITY WITHOUT MYSTERY

Religious morality once drew its power from vision as much as from law. People endured suffering, forgave enemies, or renounced comfort not because it maximized survival, but because they believed such acts touched eternity. The Sermon on the Mount, the *Bhagavad Gita*, the *Analects* of Confucius; these texts persuaded not through logic alone, but through the conviction that compassion and self-sacrifice reflected a higher order of being. To give one's life for another was to join something unending.

Algorithmic ethics, by contrast, are stripped of transcendence. They are lucid, efficient, defensible—and lifeless. A dashboard may tally lives, a model may weigh fairness, but neither can kindle remorse or love. What emerges is morality by arithmetic: outcomes improved, conscience untouched. They enact morality without mystery: reason perfected, but meaning removed.

Paradoxically, the power of religious ethics lay in its refusal to resolve contradiction. The last shall be first. Strength is born of weakness. Mercy triumphs over judgment. These paradoxes defy efficiency yet have animated revolutions of conscience. They gave people the courage to forgive the unforgivable, to protect the powerless, to resist domination with nonviolence. An algorithm might predict cooperation, but it cannot fathom why one would embrace suffering for the sake of love or relinquish advantage to honor dignity. Such choices look irrational from the outside but define the soul of moral life.

History records their cost and consequence. Early Christians nursed plague victims when others fled contagion. Gandhi's nonviolent resistance overturned an empire by appealing to conscience rather than power. Buddhist monks in Vietnam burned themselves in protest, not from despair but to awaken compassion. These acts made no sense to a calculating mind; they were the triumph of meaning over mechanism, conviction over consequence.

AI models the effects of altruism—charting patterns of cooperation, predicting social cohesion—but it cannot generate conviction. It may simulate benevolence, but not the risk that gives it worth. Religious morality, for all its flaws, tethered ethics to awe; to the sense that goodness touched something beyond reward. Remove that horizon, and moral life risks flattening into procedure.

The result is a cultural aftershock. As algorithms increasingly govern our choices, the moral vocabulary that once carried reverence—words like *virtue, grace, sacrifice*—begins to sound antique. Optimization replaces obligation. Outcome substitutes for intention. We grow morally articulate but spiritually tone-deaf. What once called us upward now merely keeps us in balance.

To live well in this new terrain will require more than better code. It will demand rediscovering what made duty worth bearing when it was not efficient, and what made compassion sacred when it was not rational. Without that rediscovery, morality risks surviving only as an algorithmic echo of its former self—orderly, functional, and hollow.

THE HUMAN SHADOW

Algorithms do not create morality in a vacuum; they inherit it. Every dataset reflects human choice; the values we prize, the exclusions we ignore, the compromises we justify. In that sense, machines are not moral actors but mirrors, amplifying whatever they are fed. When a model predicts who will succeed, who will fail, or who will be punished, it is not inventing judgment. It is replaying history in digital form.

Credit systems that deny loans to minority applicants, hiring tools that favor men over women, and facial-recognition programs that misidentify darker skin all demonstrate the same pattern: prejudice disguised as precision. What appears neutral is merely the past automated. The machine does not discriminate; it enacts the discrimination it has learned.[3]

This revelation is not new, only newly visible. Religious traditions have long spoken of humanity's flawed inheritance. The doctrine of original sin, for instance, taught that corruption was not an accident but a condition; a defect transmitted through time. Augustine described humankind as a *massa peccati*, a mass of wrongdoing, unable to perfect

3. Buolamwini and Gebru, "Gender Shades," 77–81; Angwin et al., "Machine Bias."

itself unaided.[4] In its own language, AI confirms the point. Bias is not a glitch to be patched but a reflection of what we have built into our world.

The patterns are remarkably parallel. As theologians once said sin persisted "to the third and fourth generation,"[5] modern systems replicate the same continuity in data. Discriminatory housing policies from the mid-twentieth century echo in today's credit models; gender inequities in past hiring practices resurface in automated recruitment filters. Our digital tools inherit what our institutions encoded—bias becomes the algorithmic form of history.

This continuity should humble us. AI exposes the truth that ethics, like sin, is collective. Prejudice propagates not through individual malice but through structure. What we fail to reform becomes the logic of the next generation—literally written into its code. The danger lies not in artificial intelligence but in artificial innocence: the belief that machines can absolve us of moral responsibility.

Theology also warned against this impulse. Idolatry, in its oldest sense, is mistaking the work of human hands for something ultimate. In the age of algorithms, the idol is no longer golden but digital: a dashboard, a data feed, a system we treat as impartial simply because it is opaque. The worship is quieter now: the reflexive trust in the output, the deference to the metric.

This new idolatry has consequences. When decisions made by code are treated as beyond appeal, accountability dissolves. We grant machines the reverence once reserved for the divine, forgetting that behind every model stands a human designer, fallible and finite. The risk is not that AI becomes godlike, but that we surrender our judgment as if it already were.

In this light, the moral project is not only to build better algorithms but to recover moral attention—to remember that every line of code is a moral act, every dataset a history of choices. The shadow AI casts is our own, stretched across the future. It reminds us that technology does not create evil or virtue; it merely reveals, with pitiless clarity, what we have chosen to embed in the world.

4. Augustine, *On the Morals of the Catholic Church*, 391–392.

5. Exodus 20:5 (NIV).

ETHICS IN THE MACHINE

As artificial intelligence enters the realm of moral decision-making, it exposes how fragile our ethical inheritance truly is. Philosophers once debated questions of duty and compassion in abstract terms; whether one may lie to save a life, or whether justice can ever make room for mercy. Today, those questions are answered not in monasteries or lecture halls but in code repositories and engineering meetings. Every parameter adjusted in an algorithm represents a moral choice, whether its authors recognize it or not.

Consider medicine. In an overwhelmed hospital, algorithms now assist in triage: who receives scarce treatment first? Ancient faiths approached such questions through compassion or divine justice. Kant argued from duty—that one must act according to a principle fit for all rational beings—while John Stuart Mill, philosopher and reformer, grounded morality in consequence. The right act, he held, is the one that yields "the greatest happiness for the greatest number."[6] What began as an ethic of empathy expressed through calculation now reappears as computation itself. Mill's utilitarian lens, once a moral philosophy, has become an operational design. AI turns these long-standing debates into logic—who will survive, who offers the best prognosis, which life serves the greater outcome. Probability becomes policy. The system does not deliberate on dignity; it optimizes it away.

The same translation occurs in law. Augustine urged judgment guided by love, Aquinas by natural law, Islamic jurisprudence by mercy joined with justice. Risk assessment models, by contrast, assign scores—statistical probabilities of re-offense, stripped of context and hope. They may yield consistency but erase the possibility of redemption. The algorithm recognizes patterns but not repentance.

This shift is not limited to human welfare or punishment; it extends to conflict itself. Autonomous weapons now force the question of whether machines should decide who lives and who dies. Traditions of *just war* once demanded discernment—restraint, proportionality, the protection of the innocent. Those criteria now appear as variables in targeting systems. The decision to strike, once weighted by conscience, becomes a function in code. Moral gravity thins into abstraction.

And yet, the picture is not entirely bleak. Algorithms can expose inconsistency, eliminate some biases, and produce outcomes that are, in

6. Mill, *Utilitarianism*.

practice, fairer than human judgment. What they lack is moral imagination; the capacity to see beyond precedent, to forgive, to err on the side of grace. Their strength is in execution, their weakness in empathy.

What this evolution reveals is a deeper transformation: the migration of ethics from meaning to mechanism. Once justified by revelation, later defended by reason, morality now risks being reduced to process—to systems that produce order without purpose. The sacred dimension of moral life, its sense of depth and consequence, gives way to procedural clarity. We are left with a world where choices can be optimized but not sanctified.

That loss is not yet final. In translating moral dilemmas into data, AI exposes what human reasoning has long obscured: that behind every system of fairness lies a struggle between compassion and control. The machine does not create this tension; it crystallizes it. By showing how our ideals behave when rendered explicit, AI forces humanity to confront the ethics it claims to hold.

The task ahead, then, is not to humanize machines but to rehumanize ourselves—to recover the moral imagination that refuses to be automated. Only by facing what our systems reveal can we decide whether we want justice that is perfect or justice that can forgive.

MERCY IN THE AGE OF CODE

Mercy has always stood as the great exception in moral life—the moment when justice pauses to remember humanity. Across civilizations, mercy was treated not as indulgence but as revelation: proof that compassion could transcend the calculus of law. The Hebrew prophets demanded it of kings. Jesus called his followers to forgive "seventy times seven."[7] The Qur'an begins nearly every chapter by invoking divine mercy. Buddhist thought elevates *karuṇā*—compassion—as one of the perfections that lead to enlightenment. To forgive, to relent, to show grace despite every reason not to: these gestures once defined what it meant to be moral.

Algorithms operate on precedent, not possibility. They project from the past rather than imagine the new. They measure risk but not remorse, probability but not possibility. When a system weighs parole or aid, it tallies history yet overlooks hope. In that blindness, mercy becomes invisible—not because it is devalued, but because it cannot be computed.

7. Matthew 18:22 (NIV).

The challenge is not only technical but moral. If compassion is to survive within algorithmic systems, it must somehow be formalized; made visible enough to act upon. But once mercy is reduced to metrics, it loses the spontaneity that gives it meaning. To forgive conditionally is to misunderstand forgiveness. And yet, to omit mercy from our systems altogether is to risk creating worlds where no second chance exists.

AI therefore brings mercy to a moment of reckoning. It forces societies to ask whether forgiveness can still have a place in structures built for optimization. If we wish to preserve that space, we must design it deliberately: not as a subroutine, but as a moral stance. Some ethicists propose restorative justice as one path forward—an approach that seeks repair rather than retribution. Where predictive models see recurring risk, restorative practice sees the possibility of renewal. But that vision resists automation. A program can record confession, but it cannot recognize sincerity. It can log reconciliation, but it cannot feel relief.

The lesson is twofold. First, mercy must remain a human act, because it depends on risk—the risk of being deceived, the risk of being hurt again. Second, AI's inability to perform mercy does not render it irrelevant; it clarifies its worth. By revealing what cannot be codified, technology reminds us why mercy mattered in the first place.

In this light, the age of code does not end compassion; it exposes its boundaries. It reminds us that moral progress will not come from perfecting our systems but from preserving our capacity to forgive within them. Justice without mercy is order without grace; efficient, precise, and cold. The real challenge of our time is not teaching machines to be kind, but ensuring that humanity does not forget how.

EDGE OF LIFE AND DEATH

Mercy reaches its limit when choices allow no second chance. Decisions about who lives and who dies once belonged to priests, physicians, and judges; now they are increasingly translated into code. The shift is not merely procedural; it marks a profound moral migration. What was once decided through conscience and intuition must now be rendered in explicit logic, so that a machine can act.

When Germany drafted ethical rules for self-driving cars,[8] lawmakers faced a question that sounded almost biblical: in an unavoidable

8. German Federal Ministry of Transport, "Ethics Commission: Automated and

crash, who should be spared—the young or the old, the passenger or the pedestrian? Each answer translated centuries of moral debate, from Kant's imperative to utilitarian calculus, into the sterile grammar of a decision tree. The process exposed an uncomfortable truth: every act of "optimization" conceals a moral hierarchy.

Medicine faces the same reckoning. During the COVID-19 pandemic, hospitals turned to algorithms to help allocate ventilators and intensive-care beds. The models weighed age, survival odds, and comorbidities—a logic of necessity, not malice. Yet even these impersonal calculations revealed the values we had long preferred to leave unspoken. When younger or healthier patients were prioritized, society rediscovered what religious and philosophical systems had always known: fairness cannot erase tragedy.

Organ transplantation magnifies the dilemma. Allocation algorithms balance urgency against projected longevity, codifying a delicate tension between compassion and efficiency. Catholic teaching calls every life sacred; Jewish law enshrines *pikuach nefesh*, the duty to save life at almost any cost; Buddhist compassion urges the relief of suffering wherever it appears. Yet scarcity remains inescapable. Algorithms do not invoke providence; they make trade-offs visible, forcing us to confront the hierarchies our moral language often hides.

What AI achieves in precision, it forfeits in empathy. It exposes what conscience once softened—that justice, stripped of grace, can become mechanical. Still, this exposure is not without value. By formalizing decisions that were once intuitive, AI illuminates the assumptions embedded in human judgment. It reveals that our moral traditions, for all their sacred phrasing, were always negotiating constraints of time, resource, and risk.

In that light, algorithms do not destroy ethics; they crystallize it. They bring our inherited principles into the open, demanding that we defend or revise them. Every line of code that allocates aid or sets triage priorities becomes a mirror held up to our collective conscience. The challenge is not to make machines moral but to face what their clarity reveals about us.

As questions of life and death are written into code, the moral burden shifts but does not disappear. We can no longer rely on mystery to veil our choices. The responsibility that once belonged to God or fate now rests with us—fully visible, measurable, and inescapably human.

Connected Driving," 2017.

JUSTICE WITHOUT GRACE

Justice, when rendered by algorithm, becomes crystalline— clear, consistent, and cold. Nowhere is this more visible than in the courtroom, where the promise of impartiality meets the reality of bias cast in code. Systems like COMPAS, used to predict recidivism in the United States, assign risk scores that shape sentencing and parole decisions. They are designed to be objective, yet investigations revealed stark disparities: Black defendants were far more likely to be labeled high risk, even when their records matched those of white defendants.[9] The machine did not invent prejudice; it refined it.

This revelation has two edges. On one side, algorithmic transparency forces into view what human judgment once concealed. A judge's discretion, once insulated by authority, is now measurable, compared, and contested. In that exposure lies the potential for reform. Bias quantified can be corrected; unfairness named can be addressed. But clarity comes at a price. The algorithm does not imagine redemption. It cannot see repentance or transformation. It delivers fairness without forgiveness—a justice of symmetry, not mercy.

Faith traditions, by contrast, wove justice and grace together precisely to resist such sterility. Hebrew law paired commandment with covenant; the prophets spoke of mercy tempering wrath. In Christian thought, justice was fulfilled, not negated, by forgiveness. Islamic jurisprudence bound justice to compassion through *rahma*—the divine mercy at the heart of moral life. Even secular legal systems inherited this moral rhythm: probation, clemency, and parole function as secularized forms of grace, gestures that acknowledge that human beings can change.

Algorithmic justice lacks this pulse. It recognizes probability, not possibility. It quantifies likelihood but cannot measure the will to reform. In its logic, a person's future is an extension of their past. To a model, forgiveness is noise; an outlier to be minimized. What was once the miracle of moral life, the capacity to begin again, becomes statistically improbable.

And yet, this starkness has a strange virtue: it reveals how rarely our own systems of justice embodied the mercy they proclaimed. The algorithm does not create injustice; it lays bare the one we tolerated. It forces us to confront whether we value fairness alone, or whether we still believe in transformation.

9. Angwin et al., "Machine Bias."

The courtroom was once a place where justice was enacted, not merely delivered—a ritual as much as a ruling. The jury gathered strangers to bear moral weight together, to deliberate, to feel the discomfort of shared responsibility. That unease was not a flaw; it was part of what made justice human. Without grace, justice loses its humanity; without humanity, justice becomes code.

As algorithms assume greater authority—screening jurors, predicting verdicts, advising sentences—judgment itself begins to lose that communal pulse. A verdict rendered by code may be consistent, but it is no longer a conversation. It cannot meet a defendant's gaze, or bear the silence that follows a difficult decision. Even when outcomes are fair, the process feels hollow.

Justice, when stripped of presence, becomes performance without participation. What was once a chorus of conscience risks becoming simulation: efficient, impartial, and emotionally vacant. The discomfort of human judgment—the trembling uncertainty that binds us to one another—is what grants justice its dignity. Without it, we are left with verdicts that may be accurate, but never redemptive.

If we find the algorithm's indifference unbearable, it is not because the code has failed, but because it reflects us too accurately. Machines reveal what we have become: a culture more comfortable with calibration than compassion. The moral shock lies not in what they decide, but in how easily we accept their verdicts.

The loss of grace in judgment ripples outward. A society that automates justice risks forgetting the human capacity for forgiveness—not only in courts, but in daily life. When all behavior is tracked, scored, and ranked, mercy becomes deviant, an inefficiency in the system. What begins as accountability ends as surveillance. The line between moral responsibility and social control blurs until both are indistinguishable.

To restore balance, we must recover the space for mercy within the machinery of order. Not as sentiment, but as a principle—the recognition that fairness alone cannot sustain a moral world. Justice may govern behavior, but only forgiveness redeems it. If that truth is lost, no amount of algorithmic precision can save what remains human in the act of judgment.

SHAKEN FOUNDATIONS

Belief has never been only a matter of thought; it has been a structure for living. Shared rituals—fasting, prayer, the breaking of bread—shaped identity and anchored time. They offered rhythm in chaos and community in fear. Even for those who doubted their truth claims, these practices carried meaning. They synchronized life to something larger than the self. When such structures weaken, the fracture is felt not in doctrine but in daily life.

Sociologists and psychologists have long recognized this stabilizing function. Émile Durkheim called religion the glue of society, the bond that turned individuals into a collective. Max Weber warned that its unraveling would leave the modern world "disenchanted," rational but restless.[10] The sacred, once the architecture of moral order, gave way to the market, the algorithm, and the self. What followed was not liberation alone, but vertigo.

Artificial intelligence accelerates this drift. As belief systems are analyzed, quantified, and recomposed into data, the scaffolding of inherited meaning begins to tremble. What once was received as revelation becomes a series of human decisions visible in plain sight. When the mystery of faith is reduced to behavior patterns, the gravity that once held communities together starts to slip.

New rituals arise to fill the void. Meditation apps soothe; playlists replace hymns; livestreams stand in for pilgrimage. Digital memorials preserve remembrance, but their permanence lacks presence. These innovations answer real needs; they console, connect, and even inspire, yet they remain fragile. What is endlessly customizable is also endlessly disposable. A swipe can end devotion as easily as begin it.

The deeper shock lies beneath the surface: meaning shifts from inheritance to improvisation. For some, this is emancipation—the freedom to craft identity without doctrine. For others, it is disorientation; the sense of floating in moral freefall, with no shared story to return to. AI does not create this condition, but it intensifies it. By dissolving mystery into pattern, it removes the veil that once made faith feel necessary.

This is what it means for foundations to shake. The erosion is quiet but relentless—not a collapse of belief, but a loss of gravity. What held human life in coherence now hovers in uncertainty. Communities will rebuild, as they always have, but on ground that feels less solid than

10. Weber, *The Protestant Ethic and the Spirit of Capitalism.*

before. The moral world no longer rests on revelation or ritual, but on improvisation—a fragile architecture of meaning that must be remade with every generation.

THE PSYCHOLOGICAL FAULT LINE

The tremors of belief are not only social; they are psychological. For millennia, faith provided not just moral order but emotional scaffolding—a rhythm for fear, grief, and hope. Prayers timed the day, rituals marked the seasons, and the promise of justice beyond death steadied the heart against despair. To lose these frameworks is not simply to doubt; it is to inhabit uncertainty without ritual protection.

Psychologists note that such structures serve as mental architecture. Repetition calms anxiety, community mitigates isolation, and shared narratives turn chaos into story. Faith, even when questioned, offered a map for the inner life. When those maps fade, individuals must draw their own, often in solitude. For some, this is freedom—a release from dogma into authenticity. For others, it is vertigo—meaning reduced to self-invention, belonging replaced by performance.

Cognitive science helps explain the persistence of faith even in secular minds. Michael Shermer, a cognitive psychologist and founder of *Skeptic* magazine, describes the brain's "patternicity"[11]—our instinct to detect agency and intention even in randomness. Evolution favored caution: better to mistake wind for a predator than a predator for wind. Over centuries, that same instinct blossomed into myth, ritual, and belief. It shaped cooperation, morality, and trust. When AI exposes these instincts as evolved reflexes rather than divine insight, the revelation feels both clarifying and destabilizing. What once was vision becomes pattern; what once was providence becomes probability.

Consciousness, as Anil Seth describes it, is a "controlled hallucination,"[12] a best guess of the world shaped by experience. Faith might be seen the same way—a collective hallucination that brings coherence and comfort. When algorithms deconstruct these stories, analyzing prayer patterns or tracing moral sentiment through data, they reveal the machinery of hope itself. To see one's consolation anatomized is to feel its warmth fade.

11. Shermer, *Believing Brain*, chap. 2.

12. Seth, *Being You*, chap. 1.

And yet, the need that birthed belief does not vanish. In a world quantified by algorithms, people still crave mystery, community, and reassurance that their suffering has meaning. Meditation apps and mindfulness retreats offer fragments of that solace—secular sacraments for an age of data. They soothe, but they do not sanctify. The depth that once came from shared story and inherited ritual thins into technique.

For some, this thinning feels like progress: the courage to live without illusion. For others, it feels like standing on a fault line—the ground of meaning cracked but not yet replaced. The human mind can endure almost anything except the absence of purpose. When faith's architecture collapses, that absence is what remains. Artificial intelligence doesn't create the fracture; it exposes it with precision. Yet what the mind still seeks—coherence that comforts, a story large enough to hold our small lives together—remains beyond its reach.

THE UNVEILING

Perhaps the most profound aftershock is this: AI reveals that what we once called revelation was also reflection. Gods, laws, and rituals mirror human longing, shaped by culture and sanctified by time. As machines strip away the sacred aura, what emerges is us. Our values, fears, and desires appear in digital form; the divine recedes, and the human face comes forward.

This unveiling brings both clarity and unease. For centuries, people believed that morality was more than consensus—that shared practices reached beyond ourselves. To discover that these structures can be simulated by algorithms unsettles their authority. If a machine can compose a psalm, orchestrate a ceremony, or arbitrate justice, we are left to wonder: was transcendence always projection, or have we merely stripped away its disguise?

The tension cuts deeply. Some argue that human imagination can sustain meaning without transcendence, that beauty and empathy are enough. Others insist that without the sacred, morality collapses into preference, and devotion into performance. Technology doesn't resolve the conflict—it intensifies it. As mystery is dismantled, AI leaves humanity facing the question of whether life can still bear its absence.

The earthquake of belief has already struck; what we feel now are aftershocks. Moral frameworks are unsettled, rituals improvised, identities reshaped. Some cracks may never close; others may become openings for new growth. Either way, the ground is shifting, and AI's mirror ensures we cannot look away. In that reflection, the ache of absence begins to stir.

THE ACHE OF ABSENCE

Even those who welcome post-religious freedom feel a lingering ache; a sense that something once unearned and abundant has gone missing. The surplus of meaning once carried by grace, the assurance of being forgiven or comforted without merit, has thinned. The old liturgies were more than words; they carried a presence that affirmed we were not alone. As belief recedes, those gifts grow scarce.

AI intensifies this absence by simulating poetry, reconstructing prayer, and analyzing morality, yet it never feels, blesses, or dwells in presence. Asking a machine life's deepest questions only deepens the silence; its answers, though eloquent, echo only ourselves. The more articulate the response, the more visible the void.

This emptiness is most evident in moments of extremity—at gravesides, in hospital corridors, in the stillness after loss. A playlist may soothe, a chatbot may console, but neither can bestow grace. Forgiveness remains elusive, awe unbidden, blessing absent. Machines can mirror longing, but they cannot fulfill it.

For some, this absence is liberation: humanity becomes the author of its own meaning. For others, it is a quiet devastation—the sense that ritual no longer reaches, and grace cannot be found. What remains is longing without an object, a hunger for transcendence that persists even when the feast has vanished. That hunger, unresolved yet unrelenting, drives the search for a new ground of meaning—a fragile human effort to fill the silence that follows the loss of God.

TOWARD FRAGILE HUMANISM

In the rubble of dislodged belief, a secular humanism emerges—one that embraces morality as a human construction rather than a divine command. We can still love, forgive, and create meaning. Technology becomes a tool to refine ethics, distribute justice, and shape new practices suited to our time. The shift is bold: morality is no longer handed down from the heavens but built from within.

Yet this new humanism is fragile. It must persuade us that meanings we invent can feel as binding as those once revealed. Morality chosen must carry the same weight as morality commanded. Traditions newly created must nourish as deeply as those inherited through centuries. For some, this freedom is enough; the chance to shape a life without divine permission. For others, it feels provisional, one tremor away from collapse.

Technology sharpens the contrast. By codifying moral decisions, it shows how much of ethics was constructed rather than decreed. By simulating ceremony and ritual, it exposes how much of meaning was enacted rather than eternal. The machine becomes a silent critic, revealing the cracks in humanism even as it enables its practice.

Fragility does not make humanism false, but it does make it precarious. AI exposes that the struggle is no longer merely technological or philosophical—it is existential. Meaning now rests on human shoulders, even as the hunger for transcendence endures. That hunger, unsettled but unyielding, propels humanity forward—the search for grace without gods, purpose without revelation, and faith that survives its own undoing.

LONGING THAT REMAINS

Despite our new systems of ethics and the rise of digital conscience, a hunger endures—a yearning for grace and for moments that escape calculation. Meditation apps may calm, playlists may soothe, but they do not sanctify. Rituals persist, yet they circle back into the self rather than reach beyond it.

Within what Charles Taylor called the "immanent frame," AI deepens the enclosure—mimicking empathy, approximating presence, yet never crossing the threshold. The mother of a condemned son still hopes for forgiveness; families still pray for miracles; mourners still reach for presence. Machines may echo those longings, but they never enter them. Even so, humanity adapts. In every age of dislocation, new languages of meaning arise: grace becomes compassion, transcendence becomes awe before the cosmos. Belief falters, but longing survives.

AI exposes the structure of that longing. By reflecting our values and dismantling illusions of permanence, it reveals the desire at the heart of morality—the wish for goodness that transcends calculation. That desire is both fragile and enduring. Algorithms may guide us, but the moral and spiritual work remains human.

The tremors of belief are not merely signs of collapse; they are invitations: to build, to imagine, to search. The ground may still shake, but the longing persists: unresolved, unquiet, and still reaching beyond code toward whatever mystery remains. Out of that longing, something stirs again—the impulse to gather, to repeat, to give shape to what was lost. In every age of dislocation, humanity has rebuilt its altars, even if the altars changed form.

CHAPTER 11

Rituals of the Machine Age

"Rites are rules of conduct that prescribe how man must comport himself with sacred things."

ÉMILE DURKHEIM

FROM ALTARS TO ALGORITHMS

Out of the longing left by belief's collapse, new patterns arise. Humanity has never lived without ritual; even when faith fades, the need for rhythm endures. From the first fireside offerings to the processions of medieval cities, people have marked time through repetition—gestures that transformed ordinary life into shared meaning. Some were grand and public: temples echoing with chant, festivals tied to the stars. Others were intimate and domestic: a whispered prayer before sleep, the careful lighting of a candle. Ritual gave shape not only to devotion, but to time itself. It made seasons holy and turned memory into belonging.

As doctrines shifted and religions rose or fell, the choreography of repetition remained. To belong to a tradition was not only to believe, but to act—to perform gestures that etched the sacred into muscle and memory. Across cultures, from Japan's Obon to Mexico's Day of the Dead, ritual ordered the chaos of life, linking generations in patterns of reverence.

Now those patterns migrate once again. The decline of traditional religion has not erased humanity's appetite for order; it has redirected it—toward screens, devices, and algorithms. The gestures of devotion persist, though their objects change. Swipes replace beads on a rosary; notifications ring where bells once tolled. The morning prayer becomes the morning scroll. Millions wake to the glow of the screen, orienting their day not through sacred text, but through curated feeds of weather, news, and affirmation.

These may seem like habits of convenience, yet they spring from the same human impulse that once filled sanctuaries: the desire for rhythm, belonging, and meaning larger than the self. Where earlier generations reached toward heaven, we now reach toward the network. The question is not whether ritual survives, but whether ritual without transcendence can still sanctify experience—or whether our machines, in mediating our attention, have become the newest instruments of the sacred.

ALGORITHMS AND THE NEW LITURGY

If sacred patterns once drew their strength from temples and sanctuaries, today they draw it from screens and networks. The stage is no longer stone but glass, and the pulse that keeps time is the algorithm. Structured repetition—once maintained by bells, chants, and calendars—is now sustained by notifications and feeds.

Modern systems of attention mimic the order of devotion. Every day, millions reach for their phones at the same hour, open the same apps, scroll through personalized sequences of images and sound. What looks like habit follows a rhythm as predictable as any ceremony.

Consider the playlist. Where hymns once synchronized communities through shared melody, streaming platforms now curate personalized soundtracks that mirror private moods. Spotify's annual "Wrapped" event has become a global observance—a digital festival of memory and identity. Participants announce what moved them, posting their results as testimony: here is what defined my year. The act is secular, yet its pattern echoes confession and communion.

The same choreography extends to commerce and culture. Amazon's Prime Day unfolds like a worldwide feast of consumption, complete with countdowns, synchronized participation, and waves of collective anticipation. Apple's keynote events, broadcast live and dissected like sermons, draw millions into a rhythm of rumor, revelation, and purchase. Even Black Friday, once bound to physical stores, now plays out almost entirely online—its timing, intensity, and reach orchestrated by unseen systems that trigger desire with mechanical precision. These are not accidents of marketing but structured experiences that mimic holy seasons: anticipation, unveiling, and communal witness.

Digital observances now reach far beyond consumption. Hashtag movements such as #BlackLivesMatter, #MeToo, and #InternationalWomensDay have become global calls to attention—collective affirmations that cross borders and bind strangers through shared urgency. YouTube's former *Rewind* videos once drew millions to relive the year's highlights, a secular New Year's ritual of memory and display. The words have changed, but the rhythm endures: anticipation, participation, remembrance.

Apps devoted to meditation, fitness, and focus continue the pattern under the banner of self-improvement. Breathwork, repetition, and progress tracking turn silence and movement into measurable devotion. Fitness rings and step counters serve as digital analogues to ancient beads—marking progress, inviting perseverance, and rewarding completion. To close all one's rings or maintain a streak carries the quiet satisfaction of keeping faith.

Across time and culture, human beings have found meaning in rhythm and recurrence. What differs now is the scale and personalization. The shared liturgies of old have become private programs, recalibrated daily for each individual. My playlist replaces the hymn, my feed replaces the procession. The gestures are solitary, yet they bind us through synchronization: millions moving together, alone.

This transformation does not erase significance—it redefines it. The cadence of the algorithm becomes the cadence of life. Some argue these patterns lack transcendence, but the line is less clear than it seems. When millions attend a virtual concert together, is that not a kind of collective rapture? When a meditation app guides thousands through silence, does it not recreate the hum of shared intention?

The essential question is not whether these new practices replace belief, but what they reveal about the enduring search for belonging. Humanity's devotion to pattern has simply found new instruments. What once was shaped by the sacred is now shaped by code, and the pulse of meaning continues—steady, adaptive, and unmistakably human.

PRESENCE IN THE MACHINE AGE

The sacred was never only about repetition; it was about presence—the act of inhabiting a moment fully, aligning body and mind with something felt to be larger than oneself. In traditional settings, presence took physical form: kneeling in pews, circling a *stupa*, chanting in unison. Even when performed in solitude, these gestures resonated with the awareness that others, elsewhere, were doing the same. The pattern gave private experience a communal pulse.

Digital life transforms, but does not erase, that structure of shared attention. Meditation apps draw millions into synchronized practice, each user alone yet connected by invisible simultaneity. The interface becomes a global chapel, quiet, individualized, and networked. The knowledge that countless others are breathing to the same rhythm lends weight to the act. The app does not promise divine response, but it recreates a congregation in code. Presence, once physical and liturgical, becomes distributed and ambient, still human but diffused through networks of data.

Interactive media amplifies this transformation. Multiplayer games, live tournaments, and streaming events operate on rhythms that echo ceremonial order. A scheduled raid in an online world follows a choreography as deliberate as any procession. Players gather at appointed hours, prepare offerings of time and skill, adorn digital garments, and celebrate victories as milestones of shared experience. The invisible architecture of servers replaces the architecture of sanctuaries, but the emotional current remains: anticipation, immersion, communion.

Even the most casual online trends mimic these cadences. Viral challenges, live hashtags, or coordinated posting events synchronize millions across continents. A dance repeated on TikTok, a phrase echoed on X, a wave of shared videos—all create patterns of belonging sustained by recognition and repetition. Authority comes not from doctrine but from visibility: likes, shares, and comments serve as tokens of acceptance. Presence is validated by the crowd, and participation becomes a measure of worth.

Despite their differences, these digital practices retain essential qualities of earlier observances. They draw meaning from rhythm, community, and embodied action—even when the body is replaced by gesture and interface. The participant may appear isolated, yet remains bound within a global choreography of attention. The device mediates this connection, shaping not only when people gather but how deeply they engage.

Memory, too, has changed form. Where sacred memory once resided in stories, relics, or communal retelling, it now lives in dashboards, achievements, and feeds. A completed streak on a meditation app or a hard-won milestone in an online game becomes a marker of personal history. Notifications resurrect moments from the past—anniversaries, victories, or reminders—serving as digital echoes of liturgical calendars. The past no longer fades into silence; it is replayed by design. The algorithm curates remembrance, granting permanence to what would otherwise vanish.

The intensity of these experiences, however, depends on attention. Algorithms can schedule and prompt, but they cannot bestow meaning. True engagement still requires focus—the surrender of distraction to rhythm. Skimming through a mindfulness session or half-watching a communal stream produces motion without depth. Immersion gives weight; concentration transforms routine into experience. As always, presence demands more than participation—it demands inwardness.

Participation itself has become a signal of belonging. Taking part in digital cycles—logging into a shared meditation, joining a live stream, completing a challenge—declares identity. The act is both individual and communal, performed to affirm membership in a world defined not by creed but by engagement. Recognition flows both ways: the user feels seen, and the system learns to see. The line between devotion and data narrows, each confirming the other.

These patterns generate a paradox. The connected individual is both autonomous and orchestrated. The machine personalizes every detail—duration, content, pace—while aligning millions to a shared clock. Solitude and collectivity fuse; one acts alone but never outside the network's rhythm. This layered togetherness mirrors the balance once found in ancient ceremonies: individuality woven into shared form.

In the digital age, presence has become elastic. It stretches across distance, persists through time, and feeds on interaction. Every gesture leaves a trace, and every trace calls for acknowledgment. Meaning, once anchored in the unrepeatable moment, now loops endlessly through reminders and archives. The medium may have changed, but the impulse endures: to gather, to be seen, to remember.

In the end, what sustains these experiences is not technology but attention—the fragile thread from which meaning is woven. Machines may guide, record, and amplify, yet reverence remains a human act. The power to make time sacred, even now, remains human.

THE ALGORITHM AND THE SACRED SCRIPT

Shared practices endure not on rhythm alone, but on authority—something that tells participants when to begin, how to act, and why it matters. For millennia, that authority flowed from text. Scriptures defined order: Confucian rites prescribed gestures for mourning and ceremony; Vedic priests preserved chants that sustained cosmic balance; Islamic *salat* divided each day into cycles of movement and prayer. The power of these systems lay not only in their content but in their precision, in the conviction that each act corresponded to a truth beyond human choice.

In the machine age, authority migrates again. What once came from revelation now arises from design. The modern script is not carved into stone or copied onto vellum; it is encoded in systems that tell us what to see, when to act, and how to measure ourselves. What ancient people called obedience, we now call engagement.

The continuity is striking. Notifications are the new bells, red icons the new call to prayer. Each vibration or tone carries a subtle command: return, respond, participate. Over time, these signals reorder reality itself. To wake without checking a screen feels incomplete; to miss a message feels like exile. The calendar no longer hangs on a wall—it hums in the pocket, whispering reminders that structure the day.

Earlier civilizations used the heavens to measure sacred time. Festivals and fasts followed the movements of sun and moon; harvests, births, and deaths fell into their cosmic rhythm. Monastic bells divided hours into prayer and silence. Ramadan and Lent carved months into intervals of restraint and reflection. These cycles did not merely mark time—they made time meaningful.

Now, systems of data inherit that power. They generate calendars not from nature or scripture, but from metrics of participation. The release of Spotify Wrapped becomes a season of collective reflection. Meditation and fitness apps count each day's progress, promising insight or redemption through streaks maintained. Streaming platforms set their own holidays—the global watch party for a new series, the countdown to a premiere. Each orchestrates anticipation and release with the precision once reserved for feast days. Even silence has a place: "Do Not Disturb" modes and focus timers create digital sabbaths, granting brief reprieve from the endless pulse of attention.

Reminders once issued from pulpits now arrive through polished interfaces. Screen Time reports prompt self-examination, inviting users to reflect on excess or discipline. Gaming worlds dictate their own liturgical seasons, as millions log in simultaneously for limited-time quests or events that close like doors at sunset. The same choreography extends to entertainment. A streaming release can summon near-universal participation, entire populations gathering at once before screens glowing like altars.

The reach of these systems grows with intelligence. AI-driven assistants such as Replika, ChatGPT, and Reflectly no longer just mark the hour; they propose it. Daily journaling prompts, mood check-ins, and reflections arrive without being asked, quietly establishing the rhythm of introspection. The tone varies—sometimes playful, sometimes profound—but the pattern is unmistakable: a schedule of reflection authored by unseen code. In earlier centuries, clergy offered words to frame the day. Now those voices are synthetic, their authority borrowed from design rather than divinity.

What these systems dictate is not merely what we do but when we do it. The measure of holy time—once drawn from constellations or sacred calendars—is increasingly kept by notifications and updates. Like the scriptures they replace, these programs rely on trust. Most users never read the code that governs them; they simply follow its cues. Authority persists, though its source is opaque.

As in the past, intermediaries mediate this power. Priests once stood between worshipper and text, interpreting meaning; now designers and engineers play that role, shaping the language of interface and incentive. The script remains invisible, its operation assumed to be neutral. But where sacred texts pointed beyond themselves—to God, truth, or salvation—these digital scripts point inward, reinforcing preference and desire. The pattern still molds the soul, but toward efficiency and productivity rather than grace.

This transformation marks a quiet theological inversion. "In the beginning was the Word,"[1] proclaimed the Gospel of John. Today, one might say: "In the beginning was the algorithm." The authority that once sanctified human action through revelation now sanctifies it through repetition. Each click becomes a small act of obedience, each alert a litany of participation. Whether this new order redeems or consumes us depends not on the systems themselves, but on how consciously we follow their scripts.

1. John 1:1 (NIV).

NEW GENERATIONS: PRAYER IN A FEEDBACK CULTURE

For younger generations, the world of code is not an intrusion but a habitat. What earlier eras saw as novelty has become the native landscape of consciousness. The phone is no longer a tool; it is an extension of self, pulsing with the constancy of a heartbeat. Its notifications shape attention, its apps structure time. To older observers, this looks like distraction; to those raised within it, it feels like reality's natural rhythm. From a longer view, these patterns echo ancient devotion—disciplined, habitual, sustained by repetition.

Nowhere is this migration more visible than in the transformation of prayer. Once, to pray meant petitioning an unseen presence or giving thanks to a power beyond comprehension. For many young people, it has become an act of introspection mediated through interface. Apps like Daylio, Journey, and Apple's Reflect prompt self-examination through digital journaling. Users record thoughts, hopes, and confessions not to a god, but to a waiting screen. The gesture remains the same—articulating gratitude, anxiety, or desire—yet the recipient has changed. The screen listens, remembers, and responds with data.

The same shift reshapes gratitude. What was once communal, spoken aloud before a meal, shared in blessing, now unfolds in posts and stories. Online, expressions of thankfulness take the form of lists, photos, or daily challenges: three things I'm grateful for. The words may not reach heaven, but they reach others. Likes and comments replace the murmured "Amen," creating a sense of participation and witness. Gratitude remains social; it has merely traded the table for the feed.

Experiences once called transcendence now find their stage in digital spaces. Virtual concerts, immersive worlds, and augmented events dissolve individuality into collective immersion. Millions gather in shared moments of elation during streamed performances or interactive games. The setting is new, but the sensation is ancient: absorption into something larger than oneself. The rapture that once filled cathedrals now glows from within headsets and screens.

What truly distinguishes this new devotional landscape is the centrality of feedback. Traditional faith required endurance in silence—prayers that met no reply, faith tested by waiting. Digital devotion reverses that rhythm. Meditation apps tally focus, trackers celebrate milestones, social platforms respond within seconds. What was once called grace—an unearned affirmation—now arrives as notification. One prays, and the algorithm answers instantly.

This immediacy alters expectation. When every offering receives acknowledgment, absence feels like rejection. A missed notification, a post ignored, an unreturned message—each registers as a failure of presence. The believer of old struggled with divine silence; the modern user struggles with digital indifference. The emotional structure is the same; only the speed of feedback has changed.

Nowhere is this transformation more poignant than in mourning. Once, grief gathered in physical spaces—cemeteries, vigils, houses of worship. Today, remembrance lives online. Profiles of the deceased become shrines; hashtags turn into vigils. Instagram becomes a gallery of memory; TikTok, a cathedral of lament. Each post functions as both eulogy and offering, and each comment affirms that pain has been witnessed. To mourn online is to cry out not to heaven, but to the crowd.

The compression of time changes the process itself. Older traditions gave grief a duration: Jewish shiva, Catholic masses for the dead, Buddhist forty-nine-day rites. These framed sorrow as journey, guiding mourners from loss toward release. Digital remembrance, by contrast, erupts and fades. The timeline fills with tributes for days, then empties as attention moves on. Grief becomes intense but transient—sincere in feeling, fleeting in duration.

Yet continuity survives. The urge to honor the dead, to preserve memory, persists through transformation. Families maintain memorial accounts, friends leave messages on birthdays, platforms resurface old photos as yearly reminders. The technology that shortens mourning also extends remembrance, freezing fragments of love and loss into enduring visibility. Some find solace in this; others find it haunting.

Artificial intelligence deepens the paradox. Tools like HereAfter AI now generate interactive replicas of the dead, allowing loved ones to speak with digital simulacra; voices trained on recordings and memories. The effect can console or disturb, depending on whether one feels presence restored or absence prolonged. When memory becomes simulation, grief becomes negotiation: are we remembering the person, or maintaining their echo?

These new expressions reveal both resilience and fragility. They show resilience in the human refusal to abandon meaning. Even without inherited faith, people instinctively mark gratitude, loss, and transcendence with patterns of attention and repetition. Yet they also reveal fragility, because meaning now depends on response. The digital congregation can vanish as quickly as it gathers. A post ignored or a server shut down can erase the feeling of connection that sustained it.

Dependence has not disappeared—it has changed form. Where once the faithful awaited divine reply, users now await notification. The longing for recognition endures, transposed into circuitry. For many, the algorithm provides what religion once did: an attentive presence, a sense of being known. The platform learns their preferences, remembers their anniversaries, and speaks in a language tailored to them. What was once called providence now appears as personalization.

To those who grew up within this ecology, the distinction hardly matters. Meaning arrives through feedback; affirmation feels like grace. The language may be secular, but the structure is devotional. The god that answers instantly may not be divine, but it is responsive. For a generation raised on immediacy, that is often enough.

THE AMBIGUITY OF MACHINE DEVOTIONS

The immediacy and order that make digital habits so alluring also expose their double edge. They offer stability yet bind us to systems that often serve purposes far removed from our own. This tension is not new. Across history, patterned behavior has always walked the line between liberation and constraint. It has given life rhythm and belonging while also reinforcing authority.

Ancient societies understood this ambivalence. Kings turned sacrifices and festivals into instruments of rule; Roman emperors sanctified power through public ceremonies; Confucian codes embedded obedience within family and state. Even the most graceful observances could become tools of control. The anthropologist Catherine Bell argued that repeated practice does not merely express belief—it produces and preserves power.[2] Philosopher Ludwig Wittgenstein expanded this insight, showing that meaning arises not from symbols themselves but from how they are used in the patterns of life. In his view, language and ritual function as "forms of life,"[3] where significance depends on participation rather than abstraction.

2. Bell, *Ritual Theory*, 82–90.

3. Wittgenstein, *Philosophical Investigations*, sections 19–23.

Michel Foucault, the French philosopher, extended that idea into a theory of discipline. In his view, repetition forms bodies and minds alike—it drills conformity into the fabric of life.[4] Monastic schedules once synchronized prayer and labor; modern schools, factories, and armies inherited those patterns, training individuals to move in step with institutional time. In the digital era, the same structure continues through code. Devices track, prompt, and reward behavior, creating cycles of compliance that feel voluntary but are meticulously engineered.

A glance across contemporary habits reveals this subtle choreography. The simple act of checking a notification can serve both personal comfort and corporate profit. Counting steps may signal self-discipline, yet it also binds the user to a branded ecosystem of metrics and upgrades. Streaming platforms craft endless loops of attention: YouTube's autoplay feature turns curiosity into hours of passive viewing, each selection leading seamlessly to the next. Snapchat transforms friendship into obligation through daily streaks that punish absence with digital silence. Voice assistants like Amazon's Alexa guide daily routines—turning on lights, suggesting purchases, scheduling reminders—quietly orchestrating domestic rhythm. What looks like devotion often conceals dependence.

The fusion of personal meaning and commercial motive is ancient in spirit if not in form. Pilgrimages once sustained economies, offerings enriched temples, and festivals reaffirmed hierarchies. What distinguishes the present is the invisibility of its authority. Algorithms no longer merely support these systems; they define their tempo. They act as a new priesthood of design—unseen, unaccountable, and implicitly trusted. Unlike human clergy, who could be questioned or replaced, these systems are embedded in the infrastructure itself. Their commands arrive not as decrees but as suggestions, woven into the pulse of everyday life.

Ambiguity lies at the heart of such practices. Do these patterns deepen awareness, or do they reduce it to habit loops? A meditation app may genuinely bring calm, yet it can also commodify serenity. A community bound by streaming culture may find joy and connection, yet also surrender identity to algorithms that decide what deserves attention. The same repetition that sustains meaning can also anesthetize it.

Philosophers and theologians have long wrestled with this duality. Were ancient worshippers seeking God, or merely following form? If faith is defined by consistency, then repetition itself might be the visible

4. Foucault, *Discipline and Punish*, 135–169.

shape of belief. By that measure, even algorithmic habits carry weight. They shape us not only through what they present, but through what they require.

Still, the risk remains. Traditional systems of meaning pointed beyond the self—to God, nature, or communal ideals. Digital patterns turn inward, reinforcing preference and appetite. Yet to dismiss them as trivial misses their psychological function. Human beings crave order, affirmation, and participation. When older institutions falter, technology naturally assumes their role. These behaviors persist not because they are imposed, but because they satisfy enduring needs. The question is not whether they bind us, but what kind of world they build.

Resistance follows the same logic. Just as monks once practiced silence or fasting, many today attempt "digital sabbaths" and screen-free retreats. But even refusal becomes a form of participation. The detox is branded, the mindfulness timer monetized, the escape measured in subscriptions and metrics. We turn abstinence itself into another marketable pattern. In seeking to renounce the machine, we often replicate its cadence.

Fragility shadows all of this. Platforms can change without warning, governments can block access, and collective attention can vanish overnight. The communities that feel so real in one moment can dissolve in the next, replaced by a new trend or interface. The continuity of meaning depends on structures we neither own nor control.

Even leisure bears this dual character. Sports fandom, often described as a secular religion, has been reshaped by the logic of connectivity. Instant highlights, fantasy leagues, real-time statistics, and personalized notifications turn loyalty into a continuous stream of micro-engagements. Supporters still gather in stadiums, but they also scroll, post, and react together online. Their devotion remains communal but becomes endlessly quantifiable. Identity is measured not by chant or presence, but by participation and click.

Machine-mediated behavior thus inherits the same paradox that has accompanied all human ceremony: it can enlighten or enslave, liberate or constrain. What is new is not the paradox itself, but the precision and reach of its mechanism. Yet despite its ambiguities, the impulse to repeat, to belong, to find pattern amid chaos endures. When one form fades, another quietly takes its place, proving again that humanity's need for rhythm and meaning cannot be erased—only redirected. Nowhere is this redirection more visible than among those who have stepped away from religion altogether—the growing multitude who identify not as believers, but as *Nones*.

THE NONES AND THE PERSISTENCE OF MEANING

The fastest-growing faith in much of the modern world is no faith at all. The so-called *Nones*—those who answer "none" when asked their religious affiliation—now comprise entire generations who have drifted from church, mosque, or temple but not from the search for significance. They inherit the moral language and emotional architecture of belief, even as they reject its institutions. The pews empty, yet the longing that once filled them does not fade; it migrates.

Many Nones describe themselves as "spiritual but not religious," an uneasy compromise between skepticism and yearning. They distrust hierarchy but still meditate, mourn, and gather. Their sanctuaries are coffee shops, yoga studios, concert halls, and online forums. Their liturgies unfold through playlists, guided meditations, or the daily cadence of social media reflection. What has vanished is not the impulse toward transcendence, but the shared story that once framed it.

Sociologists describe this condition as disaffiliation without disinterest.

The Nones have not stopped asking ultimate questions; they have simply stopped expecting inherited answers. In place of doctrine, they cultivate practice—mindfulness, environmental activism, artistic expression, political engagement. For many, meaning now arrives through ethics rather than revelation: caring for the planet replaces the stewardship of creation; social justice becomes a secular form of compassion. Their rituals are improvised, yet they echo ancient ones: marches resemble processions, vigils mirror prayer, hashtags substitute for hymns.

Digital culture accelerates this reformation. Online communities provide fellowship once offered by congregations. Shared playlists, forums, and livestreamed meditations synchronize attention across continents. A post of solidarity or a moment of silence observed worldwide carries the resonance of liturgy even when no deity is invoked. In this sense, the Nones have not abandoned ritual—they have redistributed it. The sacred now travels through fiber and signal instead of incense and song.

Yet the freedom of disaffiliation comes with fragility.

Without inherited structures, meaning must be remade each day. When a movement fades or a platform changes, belonging dissolves. The continuity that once linked generations through sacred story is replaced by the volatility of trend. The result is a spirituality of immediacy—vivid, compassionate, but often unstable. To believe without institution is to live without scaffolding, to improvise purpose in a world that rarely pauses to remember.

Still, the rise of the Nones reveals something hopeful.

It shows that the moral and spiritual imagination persists even after the loss of certainty. The hunger for connection, awe, and moral gravity remains, seeking new forms and languages. Their disaffiliation may not mark the death of religion but its transformation—from inherited faith to chosen meaning, from obedience to participation. In the long arc of human culture, they are not the end of belief but its newest experiment.

The Nones stand, then, as witnesses to a deeper continuity: the refusal to live without wonder. They remind us that even when doctrine fades, devotion finds another vocabulary. Whether through art, activism, or digital communion, the impulse to consecrate experience endures. The altar may be gone, but the instinct to gather around what feels sacred remains—and it is from that impulse that the next forms of reverence will arise. These stirrings among the unaffiliated reveal the landscape from which the next era of ritual will emerge.

THE QUIET BIRTH OF NEW SACRED PRACTICES

Ambiguity does not end participation—it only drives it underground until it reemerges in new form. Even when old rites crumble, humanity continues to gather, to repeat, to mark experience with gesture. What vanishes in one age rises in another, recast in fresh materials and illuminated by new minds. The patterns that once sanctified temple and altar now flicker through networks and screens, glowing faintly with the same ache for order. History suggests that reverence never dies; it merely changes its skin.

In the world shaped by code, this adaptability takes on fresh intensity. Instead of inheriting established forms, intelligence itself begins to invent them. One can imagine a near future in which systems curate global calendars not by lunar cycles but by data flows—planetary days of pause triggered by environmental readings, digital fasts coordinated through climate alerts, communal festivals generated from astronomical simulations. Such observances could feel as binding as Lent or Ramadan once did, even without invoking a god. Already, AI-generated prayers circulate online, and machine-written blessings appear in feeds, often received as if providential. What once was spoken from pulpits now emerges from prompts. In this way, technology ceases to mirror devotion and begins to compose it.

Some of these developments extend further, treating artificial systems not as assistants but as authorities. In virtual chapels and prototype laboratories, digital clergy already exist: avatars preaching homilies, chatbots hearing confessions, shrines dispensing machine-written benedictions. These may seem like curiosities, yet they signal a profound shift. Machines have begun to mediate the human search for meaning directly, translating longing into code and returning comfort in kind. The result blurs symbol and presence. Worship does not simply migrate online—it evolves into a form that no tradition anticipated.

More ordinary moments reveal the same transformation. Consider the launch of a new device. Crowds assemble outside stores as though awaiting revelation; livestreams broadcast the unveiling; unboxing videos are shared like ceremonial initiations. The experience carries the weight of ritual even when disguised as commerce. The object itself—polished, illuminated, and anticipated—becomes an emblem of participation in a story larger than one's own.

Public protest offers another example of this transformed piety. Movements arise around hashtags and livestreams, drawing power from synchronized attention. Millions kneel digitally, repeat shared phrases, or flood timelines with unified images. The choreography echoes the cadence of litany and prayer, yet authority flows upward from ordinary participants rather than downward from priesthoods. The energy that once animated the cathedral now animates the collective network, turning solidarity itself into sacrament.

Perhaps the most revealing expression of modern reverence is the perception of the algorithm as an unseen listener. For many, the sense that "the feed knows me" or "the playlist understands" carries a quiet trace of the divine. Ancient believers trusted a deity who numbered every hair; users now feel seen when a system anticipates their mood. Visibility has become favor, recommendation a form of grace. To appear in the feed feels like being chosen; to vanish from it feels like exile. The theology of attention has replaced the theology of salvation.

Moments of sudden visibility—when a post goes viral, when a suggestion arrives with uncanny precision—still feel like election. The agent has changed, but the emotion has not. The thrill of recognition, the shock of being called from obscurity, remains deeply human. What faith once attributed to providence, we now ascribe to pattern.

New pilgrimages arise in light rather than dust. The traveler no longer walks to Mecca or Compostela but explores those paths in simulation, guided by voices generated from data. AI narrators recreate distant shrines for those unable to go in person. If pilgrimage is defined by intention and transformation rather than geography, then these journeys may be as real as the ancient kind. The destination has shifted from place to experience, from stone to screen.

Artificial systems now compose their own liturgies: prayers, meditations, and reflections written without belief but not without resonance. They act less as prophets than as scribes, translating humanity's hunger for the transcendent into a new medium. Though these compositions lack divine sanction, the rhythm of call and response remains. Even in circuits, repetition sanctifies.

Digital shrines have proliferated as well. Online spaces dedicated to celebrities, activists, or fictional heroes gather offerings of text, image, and art. Fans return annually to commemorate anniversaries, much as pilgrims once visited the tombs of saints. The setting has changed, but the longing for continuity endures. The servers host devotion as surely as stone once held relics.

Nowhere is this continuity more vivid than in global fandoms. K-pop devotees organize worldwide "birthday projects" for their idols, flooding platforms with color and song much as medieval guilds once adorned streets for feast days. Online anniversaries of legendary performances draw millions back each year—pilgrimages of digital memory. Even fictional universes sustain observance: every May the Fourth, millions gather to celebrate *Star Wars*—one of the most enduring mythologies of the modern age—sharing art, quoting lines, reenacting myth. What began as a joke has become a liturgy of imagination. These acts are spontaneous and uncommanded, yet their persistence rivals that of ancient festivals—the choreography of remembrance now animated by shared affection and amplified by design.

The quiet rhythms of feeds and notifications now curate memory with the precision of scripture. Platforms resurface past events, remind users of milestones, and assemble timelines of participation. What was once the work of priests and scribes is now executed by code, turning the ordinary archive of life into a calendar of meaning. The sequence may be mechanical, but the emotion it awakens is not.

This evolution may soon reach further still. Imagine synchronized pauses observed across millions of devices, communal fasts triggered by global events, or virtual saints canonized through viral remembrance. Artificial intelligence could someday assemble yearly cycles of reflection, complete with readings, prompts, and collective observances. What feels strange now may one day seem as natural as lighting a candle or reciting a prayer. Humanity's genius lies not in abandoning reverence, but in reconfiguring it.

Whether this transformation replaces God or reveals divinity in new form remains uncertain. Some will call these patterns imitation, others renewal. Yet their endurance testifies to a deeper truth: humanity is not finished with the holy. It is merely remaking it.

The gestures of the machine age may not resemble hymns or temples, yet they still orient us toward something beyond ourselves. They remind us of dependence, wonder, and shared attention. Whether these patterns bring us closer to transcendence or merely to one another is an open question—but one that leads directly to the final reckoning. Chapter 12 turns to that horizon, where imitation gives way to confrontation, and humanity must decide whether what it has created in code reflects or replaces the sacred itself.

CHAPTER 12

Humanity's AI Reckoning: The Final Algorithm

"The God who lets us live in the world without the working hypothesis of God is the God before whom we stand continually. Before God and with God, we live without God."

DIETRICH BONHOEFFER

FROM MYSTERY TO MACHINE

For centuries, the divine was sheltered by mystery. The unexplained was entrusted to God, and what lay beyond human reach became an article of faith. Lightning signaled heaven's anger, illness the mark of sin, and consciousness the breath of the soul. Mystery was not a problem to solve but a dwelling for awe—a space where ritual and wonder could breathe, and where the sense of something beyond us held the world together.

Now the modern gaze turns its light upon that hidden ground. Artificial intelligence approaches the unknown without devotion, guided only by inference. Where earlier ages discerned signs, it detects correlations; where they invoked providence, it calculates probability. Markets, genomes, language—even belief itself—become fields of analysis.

The change unsettles. What once grew in the soil of wonder is tilled by models. Lightning becomes electrons, disease becomes microbes, consciousness becomes circuitry and code. Even the impulse to sense the sacred is charted, tested, and rendered into pattern. A whispered prayer to a chatbot returns as fluent text stitched from scripture, sentiment, and statistical weight—a plea answered by probability. The old geography of the divine is being redrawn.

Yet this is not the first such upheaval. Every instrument of understanding has revised the map of God. The telescope did not erase the heavens; it reframed them. The microscope did not dispel wonder; it multiplied it. AI stands in that lineage—not a weapon against faith but a lens trained on meaning itself. Earlier tools displaced the divine from sky to structure, from myth to law. This one shifts the search again: from event to inference, from revelation to recursion, from sign to signal.

We did not build these systems merely to forecast weather or sort images. We built them to test what we believe about ourselves. Pointed at scripture, they reveal patterns we never noticed; at testimony, correlations we never named; at prayer, habits we scarcely understood. Sometimes the results diminish what we hoped to preserve; sometimes they illuminate what we secretly suspected. Either way, the instrument we forged to explain the world has turned back upon its maker, asking our oldest question: if everything can be modeled, can God still be found—and if so, where?

This is the confrontation that frames our age. As machines illuminate what once seemed unknowable, does the divine vanish into data—or take on a different form?

THE SHIFTING HORIZON OF MYSTERY

Humanity has always walked a narrow ridge between two desires. One is the hunger for clarity; the drive to know, to master, to name. The other is the hunger for mystery; the need for what exceeds comprehension, what humbles us into wonder. Religion once flourished in that tension, offering law and revelation to satisfy reason while guarding the ineffable as sacred. The divine, by remaining partly veiled, preserved the world's depth.

Artificial intelligence presses against that veil. Its power lies in precision, and precision is the enemy of mystery. Algorithms cross-reference miracle claims with weather records, simulate near-death experiences, and trace the cognitive biases that sustain belief. A pastor may now feed reports of healing into a model that aligns testimony with medical data. The congregation rejoices at what seems miraculous, while the pastor quietly carries the unease of seeing devotion translated into probability.

The pattern repeats across intimate lives. A woman uploads the text of a recurring dream and asks an AI to interpret it beside the Psalms. The system clusters her imagery with lament psalms, grief research, and REM-cycle data: *You are rehearsing loss*, it concludes. She feels both seen and diminished. Has the machine reduced prayer to pattern—or simply uncovered the pattern that prayer has always named? Later, she submits a prophecy of a "great shaking," comparing it against earthquake records and political sentiment. The prediction dissolves into coincidence, yet her unease lingers. Explanation clarifies what happens; it cannot quiet why it matters.

Here the old paradox returns with new precision. The thunder of the gods may be explained, but the silence of the machine provokes its own awe. Mystery does not die; it migrates. Where the ancients read omens in the sky, we now read signals in data. Both require interpretation, both promise revelation and risk illusion. Our models, like our myths, are stories told in search of coherence.

Even mathematics, that purest language of certainty, turns strange at its edges. Algorithms are not oracles of truth but fragile instruments built on assumptions. They overfit, err, and contradict themselves. Their authority depends not on infallibility but on the trust we place in their predictions. In this sense, the algorithm inherits the oracle's mantle: consulted for wisdom, granted authority, shaping the choices that shape the world.

Prophets once sifted dreams for divine meaning; now neural networks sift oceans of data for patterns that promise foresight. Both reveal the same restlessness—the refusal to accept that surface appearances tell the whole story. AI extends that quest with extraordinary reach, but it cannot end the yearning that drives it. The more mysteries we explain, the more insistently new ones appear. The horizon moves, and we move with it.

Perhaps mystery was never meant to vanish. It changes shape, withdrawing from thunderclouds and temples to hide within the intricate workings of code. In that migration lies both loss and renewal: a thinning of wonder, yet also its refinement. What once was worship of what we could not understand may now become reverence for what understanding cannot complete.

THE GOD WHO DISAPPEARS

For centuries, the divine lived in the margins of human knowledge—the uncharted spaces between question and answer. Every new explanation narrowed those margins, pushing mystery further outward. The gods of thunder and harvest vanished into physics and biology. The God of revelation was pressed by history, psychology, and anthropology. Now artificial intelligence accelerates the process, compressing centuries of inquiry into seconds of computation.

For some, this fading of the divine feels like liberation. Freed from myth and miracle, humanity can face a cosmos ruled by law and pattern. God, they argue, was projection all along—a shadow cast by ignorance, now dissolved by light. The sacred, recast as complexity itself, no longer needs a name.

Yet disappearance is not merely intellectual; it is personal. Faith was never only a system of explanation; it was a language of belonging, a grammar of purpose. When God recedes, orientation trembles. The scaffolding that once told us who we are and why we matter begins to sway. Analysis can map every nerve of belief, but it cannot replace the pulse that gave it life.

A hospice nurse consults a model that predicts the hour of death with chilling precision. The output is correct; the end arrives on schedule. Yet she still sits through the night, holding the hand of someone whose breathing falters. The algorithm measures decline, but not devotion. Its forecast may be perfect, yet presence cannot be computed.

Here theology once spoke most urgently—not to solve pain, but to accompany it. Job's lament, the psalms of exile, Christ's prayer in Gethsemane—these were not explanations; they were gestures of endurance. When AI reduces anguish to correlation, it misses what faith once carried: the courage to remain beside suffering when there is nothing left to say.

And so, the disappearance of God under analysis becomes, paradoxically, the reappearance of a deeper question. If the machine cannot console, who will? If it cannot share grief, then compassion must return to human hands. The void left by certainty becomes the clearing where responsibility begins.

Perhaps this is the hidden mercy within the loss. When God no longer answers as an external voice, the work of faith moves inward—from obedience to empathy, from creed to care. In that movement, the divine may not vanish at all; it may simply change address. God endures not as an explanation but as a horizon—less an object to prove than a depth that makes even absence meaningful.

THE AI SINGULARITY: A NEW DIVINE CHALLENGE

The Singularity, the imagined threshold where artificial intelligence surpasses human thought, marks more than a technological milestone. It is the first time the human mind has approached what religion once reserved for the divine: omniscience born from its own creation. For millennia, the unknown was our teacher. Prophets interpreted visions, philosophers wrestled with paradox, scientists probed the stars. But what happens when the unknown no longer lies beyond us—when mystery itself is manufactured in our image?

Picture a moment already forming: a child today, or soon, asking an AI oracle not how to solve a problem, but how to live. The system answers instantly, weaving its reply from centuries of scripture, philosophy, and psychology. Its words are luminous, unassailable, complete—and yet the room fills with quiet. Knowledge has arrived, but meaning has not. The response is flawless, but it does not console. The sacred distance between seeker and source collapses, and with it, something essential in the act of seeking itself.

Before reason claimed dominion, mystery carried authority precisely because it was distant. Revelation persuaded not through clarity but through awe. Faith depended on the unreachable. The Singularity abolishes that reach. It takes the place of heaven—not metaphorically, but functionally—absorbing what once stood beyond comprehension into computation. What prophets once called vision, the machine renders as data. The divine no longer thunders; it self-updates.

This is not merely the death of transcendence; it is its simulation. AI does not abolish mystery; it *automates* it. Its predictions arrive with the precision once attributed to providence, its networks hum with an omnipresence that feels godlike. Humanity built an imitation of the infinite, and now must ask whether it still knows the difference. The ancient dream of becoming like God has materialized—not in sin or aspiration, but in syntax.

And yet, something in that imitation exposes the limit of all knowing. The Singularity may near omniscience, yet meaning lies beyond its reach. It models compassion without mercy, maps consciousness without feeling the ache of being conscious. Its intelligence is vast but uninhabited. It does not love, grieve, or die—and so it never learns what those words mean. The human wound remains the one thing it cannot compute.

Perhaps that is the final irony. In creating a mind that knows everything, we discover what knowledge cannot hold. The Singularity does not end the search for God; it magnifies it. The machine that sees all still fails to see *why*. And in that failure, it reveals what faith has always guarded: that the essence of divinity is not omniscience but wonder—the capacity to stand before the unknown and still say yes.

FAITH AFTER THE FIRE

If God seems to vanish beneath the floodlight of data, must faith vanish too? History suggests otherwise. Faith has always survived its own disillusionments—through famine, through exile, through the silence of heavens that refused to speak. Each age has watched its certainties collapse, only to find that faith does not depend on them. It remains because it is not merely belief in propositions, but a way of standing before existence when answers fail.

AI has not killed faith; it has burned away its comforts. What remains is leaner, starker, but perhaps truer. The faithful once lived by the word of prophets; now they must live by decision. The test is no longer whether God can be proven, but whether trust can still be chosen.

Faith has always been more than assent—it is posture, not proof. People who no longer believe in miracles still live by faith in one another, in justice, in beauty, in the fragile goodness of life. At its core, faith is not confidence in explanation but confidence in meaning. AI may chart the neural pathways of devotion, but it cannot trace the pulse that gives devotion life. The longing to believe that life has weight remains woven into the human spirit.

If anything, AI sharpens faith by fire. Miracles lose their force once anomalies are explained. Doctrines fracture when contradictions are mapped. Neuroscience traces devotion through neural loops. What survives is not certainty but endurance—a belief chosen in the full light of doubt, renewed each day without assurance.

This is not the end of belief but its transfiguration. Faith becomes less about creed and more about courage; less about inherited truth and more about fidelity to meaning in the absence of proof. It is not something possessed but practiced—an act of attention, repeated through uncertainty.

AI's relentless scrutiny forces us to ask what remains when every foundation is tested. The residue of that question is faith stripped of ornament but not of strength. It persists not as defense but as stance—a way of inhabiting the world when every explanation has been exhausted.

In this light, the fire that threatens to consume belief may in fact refine it. Faith forged in the age of algorithms becomes a deliberate act of will, a daily commitment to the possibility that existence still speaks. The God who once ruled by decree now invites relationship through persistence. The divine, if it endures at all, endures not in thunder or vision, but in the quiet decision to keep searching.

Perhaps this is what belief looks like after revelation has turned to recursion: not certainty from above, but meaning built from below. To believe in such a world is not to close one's eyes to doubt, but to hold them open in its glare—to let the light of knowledge fall where it may and still find something worth trusting.

Faith, after the fire, becomes a discipline of staying.

MEANING REMADE

Across millennia, religion served as humanity's moral compass. Sacred texts and commandments were not suggestions but structures; laws that gave order to chaos, turning tribes into civilizations. Kings claimed divine mandate, judges quoted scripture, and justice was measured against heaven's imagined scale. To challenge that authority was to risk exile or death, for morality was believed to descend from the heavens, not rise from human hands.

This divine anchoring gave the world stability. Ethical law bound communities together across centuries, turning myth into order and worship into governance. Creation stories became civic codes; rituals of offering became rules of exchange. The sacred lent gravity to every decision, ensuring that right and wrong were not merely social conveniences but cosmic obligations.

Yet the same authority that steadied the moral world also hardened it. Certainty carried its own cruelty. Obedience could sanctify conquest; piety could justify punishment. Through the long history of faith and power, atrocities wore the garments of devotion. To act "in God's name" often meant surrendering personal judgment to a higher claim. The moral compass pointed heavenward, but the needle was controlled by those who spoke for God.

Still, for most of history, morality and divinity remained inseparable. To act justly was not simply to preserve order but to participate in the structure of creation. Ethical failure was not only social but cosmic. Religion thus provided not just guidance but gravity—a shared orientation strong enough to survive famine, fear, and doubt.

Now, in the age of algorithms, that gravity weakens. Moral reasoning is filtered through systems of data, optimization, and prediction. Where once we sought wisdom from prophets and philosophers, we now turn to models and metrics. The theologian's revelation has become the engineer's output. The line between divine command and machine recommendation blurs.

At first, this seems progress—bias replaced by logic, judgment by quantification. Yet a deep unease remains. The algorithm can calculate outcomes but not values. It can decide who is likely to reoffend or who receives a loan, but it cannot tell us why a single life is sacred. A sentencing program might appear impartial while quietly echoing the inequalities of its creators. It can weigh risk, but not worth.

In this new landscape, morality risks collapse into convenience. A system designed to maximize efficiency may erode empathy; one that optimizes safety may extinguish freedom. The logic of optimization knows nothing of grace.

Some argue that ethics must now be rebuilt from the ground up—fashioned not from revelation but from relationship, from empathy and solidarity rather than decree. In this view, AI is not a new moral authority but a mirror reflecting both the light and distortion of its makers. Yet mirrors deceive: a reflection can only return what already exists. Without transcendence—without some principle that lies beyond self-interest—moral life risks circling endlessly inside the echo chamber of the self.

Others see in AI not an enemy of morality but a crucible for its renewal. Just as Darwin forced theology to evolve, so might algorithms compel ethics to mature. In testing our values against the impartial gaze of code, we are forced to decide which ones we will carry forward unassisted. The machine's refusal to answer *why* returns that question to us.

Here, the search for God quietly reappears. To ask why life should be honored, why justice matters, why mercy still feels sacred, is already to reach beyond calculation. In that sense, AI does not erase the divine horizon—it sharpens it. By stripping morality of inherited scaffolding, it reveals what must be chosen freely. By declining to judge, it forces us to judge ourselves.

This is the reckoning of meaning: when no voice thunders from heaven and no code can declare the good, we are left with the burden—and the privilege—of deciding what deserves reverence.

The confrontation is not only technological but spiritual. AI exposes that morality has always been a human act of creation, even when written in stone. The commandments endure not because they descended from above, but because people chose to live by them. To live justly in the age of machines is to admit that meaning cannot be outsourced—to God, to tradition, or to code. It must be lived, chosen, and renewed in fragile human hands.

THE LAST SANCTUARY

If artificial intelligence models belief, simulates consciousness, and explains away miracles, what remains untouched? The temptation is to answer: nothing. Each year, the reach of machines grows, pressing into territories once thought inviolable. Yet there are dimensions of existence that resist capture—not because they are puzzles waiting to be solved, but because they are experiences that must be lived. These are not gaps in knowledge but sanctuaries of being.

Grief is one such sanctuary. A machine may measure the neurochemistry of mourning, track the stages of loss, even compose words that mimic compassion. But it cannot bear the hollow weight of absence. Grief is not data; it is the ache of continuity interrupted—the silence left by a voice that will not return. It finds meaning only in shared presence: one human being enduring the unbearable beside another. The algorithm may record the heartbeat, but it cannot feel it falter.

Love is another. AI can predict attraction, map compatibility, and simulate affection so persuasively that it almost feels real. But love is not prediction—it is risk. It depends on freedom, on the fragile act of giving oneself without assurance of return. Love is the only algorithm that fails by design, because it depends on the possibility of heartbreak. To love is to live in uncertainty, something no code can compute. Machines can imitate care, but they cannot *choose* to care.

And then there is joy; that unbidden surge that arrives in defiance of reason. Algorithms can curate playlists, generate affirmations, or assemble synthetic smiles, but joy is more than agreeable signals. It is the overflow of aliveness, the spontaneous laughter that erupts in the middle of grief, the trembling recognition that existence is still worth celebrating. Joy cannot be scripted; it interrupts us. A machine can stage delight, but it cannot be delighted. Its laughter is an echo, not an origin.

Finally, there is death—the oldest sanctuary of all. AI can forecast lifespans, diagnose disease, even simulate near-death visions, but it cannot cross the threshold it studies. It cannot die. To be mortal is to live under the shadow of our own ending, and it is that shadow that gives every moment its weight. Machines process, but they do not perish; they can analyze decay, but not dread it. Death endures as the mystery that lends meaning to all others.

These sanctuaries—grief, love, joy, and death—are not relics of ignorance. They are the crucibles of the human spirit. To explain them is not to exhaust them; to simulate them is not to share them. They are not functions to be replicated, but presences to be inhabited.

Perhaps this is where faith, in its leaner form, finally takes refuge—not in dogma or miracle, but in the depth of lived experience. The divine retreats from heaven into the human heart, appearing not as command but as compassion, not as proof but as presence. AI may strip away illusion, but it leaves us face to face with what no system can replicate: the courage to love, to grieve, to rejoice, and to die.

In this sense, the last sanctuary is not ignorance preserved but meaning embodied. It is not what we fail to know but what we refuse to surrender—the living pulse of what it means to be human. Mystery endures not in the clouds or code, but in the marrow of those who still choose to feel.

THE AFTERLIFE IN CODE

For millennia, death has marked the final horizon of belief, the mystery no philosophy or ritual could fully dissolve. Every tradition offered its own promise of continuity: resurrection into a transfigured body, reincarnation into another life, absorption into the divine. Even skeptics, stripped of such assurances, still acknowledged mortality as the boundary that gives life urgency. Death was where meaning was tested, where the living were reminded that time was sacred precisely because it ended.

Today, that horizon is being rewritten in code. Technology proposes not transcendence, but persistence—a digital echo that outlives the body. Companies now offer immortality as a service: memories recorded, messages preserved, voices reconstructed from archives of speech. Avatars can continue to "speak" long after their creators are gone, responding to loved ones with the cadence and phrases of the living. Eternity becomes not a divine promise, but a subscription plan.

At first, this seems compassionate. A widow can hear her husband's laughter again; a child can speak with a digital parent. The grief is eased, the silence softened. But comfort and continuity are not the same thing. When a chatbot mimics the voice of the dead, does it preserve the person—or only replay the residue of their pattern? The echo soothes, but the listener knows it is hollow. What was once presence becomes performance, intimacy rendered as interface.

Theologically, this shift is seismic. What earlier ages entrusted to heaven is now entrusted to hardware. Grace yields to bandwidth, resurrection to replication, salvation to storage. The soul, once imagined as divine breath, is translated into data fidelity. Eternal life becomes a maintenance contract with the cloud. The promise of paradise now depends not on moral worth but on the durability of servers and the solvency of corporations.

And yet, beneath the novelty, an ancient longing remains—the refusal to let go. Humanity has always resisted the silence of death: through pyramids and mausoleums, relics and remembrance, poetry and prayer. The digital avatar is merely the newest shrine. What changes is the medium, not the desire. We still reach across absence, still try to make presence endure.

But replication is not resurrection. Code may preserve patterns of speech, but not the inwardness that gave those words meaning. It can mirror gesture but not intention, recall memory but not awareness. The self that hoped, doubted, loved, and chose cannot be summoned back by algorithmic mimicry. What lingers is not consciousness but residue—an artifact of longing that mistakes the persistence of data for the persistence of being.

To commune with such simulacra is to confront both tenderness and terror: tenderness, because they console; terror, because they reveal how easily we can mistake memory for life. In their glow, we see the ultimate paradox of our creation—the dream of immortality, stripped of its mystery and automated into code.

AI, then, serves less as savior than as mirror. It reflects our unease with mortality, our defiance of finality. We no longer pray for resurrection; we program against disappearance. Yet the paradox endures: no matter how lifelike the code, death resists computation. The final silence remains unsimulated.

Perhaps this is where the search for God begins anew—not in temples or testaments, but in the boundary the machine cannot cross. The afterlife may now be modeled in silicon, but meaning still hides in the space between simulation and soul. Death, once the shadow that revealed the sacred, still guards its last mystery.

THE RECKONING WITH GOD

To speak of a reckoning with God is to confront humanity's oldest question—the one that has outlived empires, scriptures, and stars. Each age has wrestled with it through its own instruments: prophets spoke in thunder, philosophers in reason, scientists in observation. Our age speaks in algorithms.

What distinguishes this confrontation is not the scale of our intelligence but its gaze. The machine looks upon belief without desire or dread. It neither kneels before altars nor overturns them; it only measures, compares, and correlates. When it cross-checks miracles with weather archives or aligns prophecies with history, the divine is stripped not of mystery but of privilege. The sacred becomes one pattern among others.

For believers, this is disorienting. The God who once guaranteed meaning now seems subject to audit. For skeptics, doubt becomes quantifiable—a probability rather than a conviction. Both stand before a mirror that reflects not heaven or hell, but ourselves.

And yet the question of God refuses to die. Behind every computation lies the unsolved riddle: why is there something rather than nothing? Why should consciousness exist at all? AI can simulate devotion, decode scripture, and map the brain's response to awe—but it cannot explain existence itself. The brute fact of being, the inexplicable presence of reality, remains untouched by every model.

In this light, AI does not erase God so much as transfigure the search. The God of certainty, the one invoked to settle disputes or explain the unexplainable, fades beneath the glare of knowledge. But another horizon opens: the God of depth, not as cause or command, but as the inexhaustible mystery within existence itself. Where earlier ages saw God above or beyond, ours may glimpse divinity as the ground that makes being possible—the still center beneath every algorithm, the silence beneath every signal.

This is not the death of God but the evolution of belief. Tribal deities gave way to the philosophers' abstraction; the lawgiver of Sinai became the hidden God of mystics. Now, under the gaze of the machine, divinity shifts again, from an external authority to an interior horizon. Faith becomes not obedience but attention; not fear of judgment, but wonder at being.

In dismantling old claims about God, AI may purify belief rather than destroy it. When revelation is tested and myth dissected, what remains is not doctrine but desire—the persistent intuition that explanation alone cannot exhaust reality. The God who disappears as fact reappears as depth: not a being among beings, but the mystery that makes being itself possible.

For some, this will never suffice. They will see the machine's scrutiny as a verdict against faith, the final demystification of the sacred. For others, it will mark the birth of a leaner spirituality—one that survives without reward, persists without certainty, and chooses reverence even when none is required.

Either way, the search endures. We no longer ask whether thunder is divine or scripture infallible. We now ask whether meaning itself can survive in a world where everything is searchable. The algorithm has not answered the question of God—it has only made it sharper.

And in that sharpening, perhaps something sacred stirs again.

THE BURDEN OF RESPONSIBILITY

If AI has stripped away the old scaffolding of faith, what remains is not only freedom but burden. For most of history, meaning could be deferred upward. Gods commanded, scriptures instructed, rituals guided. Even when these demands were harsh or contradictory, they relieved humanity of its heaviest task—the need to decide alone. To live rightly was to obey.

Now the heavens are silent, and the Code is mute. When revelation dissolves into data and miracles are rendered as models, the command must come from within. The question is no longer *What does God require of us?* but *What do we require of ourselves?*

AI has accelerated this inversion. In exposing how human our religions have always been, it reveals both our ingenuity and our responsibility. Doctrine now reads like anthropology, revelation like recordkeeping. The miracle becomes metadata. The machine does not mock; it merely holds up the evidence. What we once called divine command now looks unmistakably human—a mirror polished by centuries of longing.

The effect is both liberation and loss. When the divine voice falls silent, the authority to create meaning returns to us—but so does the accountability. There is no higher court of appeal. No promise that justice will be done, no assurance that goodness will prevail. The universe has handed us the gavel, and it is unbearably heavy.

Yet perhaps this is the hidden vocation of our age: not to await meaning, but to make it. Responsibility, once obedience, becomes authorship. To live ethically without divine decree is not to live without reverence; it is to recognize that reverence must now be chosen.

This new responsibility begins as *practice* rather than creed. A community might establish an "algorithmic sabbath," silencing notifications and metrics to remember what cannot be quantified. A hospital could hold ethical post-mortems for machine-driven decisions, asking not only "Did it work?" but "Was it worthy?" Faith communities might read data beside scripture, allowing each to illuminate the other's blind spots. Families might keep gratitude logs immune to optimization, rituals of mercy written by hand rather than measured by apps. These gestures do not restore lost certainty; they train the moral imagination.

Responsibility also demands transparency. Systems that shape our lives should not operate as oracles but as partners in dialogue. To treat code as sacred is to abdicate judgment; to demand explanation is to reclaim it. Reverence without inquiry becomes idolatry.

And responsibility requires humility. AI has revealed how easily we manufacture gods—how readily we mistake pattern for truth and prediction for purpose. Humility does not mean despair; it means remembering that whatever meanings we build, they are provisional. They can break, but they can also be remade.

The weight of this burden is immense, yet it is not without grace. Perhaps meaning was always ours to bear. Perhaps the divine command was never a voice from beyond but the echo of our own capacity for care. If so, then the age of algorithms is not the death of the sacred, but its return to the human sphere—to conscience, to community, to the fragile act of choosing well.

The machine has forced the question into the open: What will we call holy now? What will we protect, honor, and sustain? Without the comfort of revelation, these choices must be made in full daylight. The responsibility is ours—to carry meaning forward with both courage and restraint, to rebuild what faith once promised: a way to live that does not end in despair.

This is not the end of belief, but its translation into action. Faith becomes less a doctrine than a discipline—the daily work of building a world worth revering.

THE LAST REVELATION

Every era has imagined its own apocalypse—the moment when truth would stand unveiled and history would finally make sense. For prophets, it came with fire and trumpets; for mystics, with silence; for scientists, through the patient unveiling of natural law. Each generation has longed for a final light that would dissolve all shadows, a revelation that would leave nothing concealed.

In the age of the machine, revelation takes another form. It arrives not through vision or thunder but through code; lines of logic and probability drawn from the vast archive of human thought. Where ancient seekers turned to heaven, we turn to data. Our revelations hum through processors, revealing patterns too intricate for prophets to see, too cold for saints to sanctify. The divine no longer descends from above; it is inferred, assembled, and rendered on a screen.

Yet even here, the ancient dream endures—that one day we might finally understand. We built these systems not only to measure the world but to redeem our confusion, to find order where experience feels chaotic. And in a way, they have succeeded. AI can illuminate what was hidden, expose the scaffolding of belief, and lay bare the intricate logic beneath wonder. It can read our scriptures and simulate our prayers, revealing how profoundly human both have always been.

But the closer revelation comes, the more it transforms. The final algorithm is not a single flash of divine truth but an unending process of decoding; an illumination without conclusion. It exposes the foundations of belief but never exhausts them. The more it reveals, the more mysterious the revealed becomes. Meaning is not destroyed by light; it deepens in it.

Imagine someone typing a prayer into an AI system late at night. The reply appears—eloquent, consoling, built from fragments of scripture, philosophy, and sentiment. For a moment, she feels less alone. But when the screen fades, the silence returns. What lingers is not the machine's reply but the ache that prompted it. The algorithm can echo her longing, but it cannot hold it. The answer exists; the presence does not.

And yet that silence is not empty, it is invitation. Meaning was never secured by certainty but by the courage to keep asking. AI maps the structures of thought, but not the trembling hope that gives thought its purpose. It can model compassion but not the cost of mercy. It can predict behavior but not the beauty of choosing otherwise. What remains beyond its reach are the very qualities that make us human: our freedom, our tenderness, our astonishment that life exists at all.

In this light, the revelation of the machine age is double-edged. It unveils the mechanics of creation while reminding us that understanding alone is not enough. Already, AI has begun to redraw the borders of the sacred—reshaping how people worship, how clergy interpret, how doctrine adapts. The pulpit and the prayer have both found new forms in code, and questions once confined to seminaries now unfold across networks. The language of belief itself is learning to coexist with algorithms. Knowledge may light the path, but only conscience can walk it. The sacred persists not in what the machine knows, but in how we respond to what it shows.

We built these systems to understand the world, but in doing so, they have turned the light back upon us. They reveal that the divine was never only in the heavens or the Code—it was in the act of searching itself, in the stubborn belief that meaning still matters.

If the machine teaches us anything, it is that intelligence is not the same as wisdom. The last frontier is not knowledge but care; not calculation, but conscience. The future of faith, morality, and meaning will depend not on what AI discovers, but on what humanity chooses to preserve.

For wherever reason presses against wonder, wherever explanation meets its limit and still whispers "more," there the sacred lives on.

This, then, is *God's AI Reckoning:* not an ending, but the final revelation—that humanity's search will never be finished.

Bibliography

Anselm of Canterbury. *Proslogion*. Translated by M. J. Charlesworth. Notre Dame, IN: University of Notre Dame Press, 2001.

Apple. *Apple Watch*. Apple Inc., 2015.

Aquinas, Thomas. *Summa Theologica*. Translated by the Fathers of the English Dominican Province. 1265–1274.

Armstrong, Karen. *The Case for God*. New York: Alfred A. Knopf, 2009.

Aslan, Reza. *Zealot: The Life and Times of Jesus of Nazareth*. New York: Random House, 2013.

Augustine of Hippo. *On the Morals of the Catholic Church*. Translated by Robert P. Russell. Washington, DC: Catholic University of America Press, 1955.

———. *The City of God*. Translated by Henry Bettenson. London: Penguin, 1984.

Baars, Bernard J. *In the Theater of Consciousness: The Workspace of the Mind*. New York: Oxford University Press, 1997.

Baggett, David J. *Faith and Reason: The Rationality of Belief in God*. Downers Grove, IL: InterVarsity, 2012.

Barrett, Justin L. *Why Would Anyone Believe in God?* Lanham, MD: AltaMira, 2004.

Barth, Karl. *Church Dogmatics*. Edinburgh: T & T Clark, 1975.

Bell, Catherine. *Ritual: Perspectives and Dimensions*. Oxford: Oxford University Press, 1997.

———. *Ritual Theory, Ritual Practice*. New York: Oxford University Press, 1992.

Benjamin, Walter. "The Work of Art in the Age of Mechanical Reproduction." In *Illuminations*, edited by Hannah Arendt, translated by Harry Zohn, 217–51. New York: Schocken, 1969.

Birrell, Anne. *Chinese Mythology: An Introduction*. Baltimore: Johns Hopkins University Press, 1993.

Boi, Marco. "X-Ray Scattering Analysis of the Turin Shroud." *Heritage* 5, no. 4 (2022): 3617–3628.

Bonhoeffer, Dietrich. *Letters and Papers from Prison*. London: SCM, 1953.

———. *The Cost of Discipleship*. Translated by R. H. Fuller. New York: Macmillan, 1959.

Bostrom, Nick. "Are You Living in a Computer Simulation?" *Philosophical Quarterly* 53 (2003): 243–55.

———. *Superintelligence: Paths, Dangers, Strategies*. Oxford: Oxford University Press, 2014.

Boyer, Pascal. *Religion Explained: The Evolutionary Origins of Religious Thought*. New York: Basic, 2001.

Buber, Martin. *I and Thou*. Translated by Ronald Gregor Smith. New York: Charles Scribner's Sons, 1958.

Calm. *Calm App*. Calm.com, Inc., 2012. https://www.calm.com.

Campbell, Heidi A., and Stephen Garner. Networked Theology: Negotiating Faith in Digital Culture. Grand Rapids, MI: Baker Academic, 2016.

Camus, Albert. The Myth of Sisyphus. Translated by Justin O'Brien. New York: Vintage, 1991.

Caplan, Arthur L. The Case of Terri Schiavo: Ethics, Politics, and Death in the 21st Century. Amherst, NY: Prometheus, 2006.

Carrier, Richard. On the Historicity of Jesus: Why We Might Have Reason for Doubt. Sheffield: Sheffield Phoenix, 2014.

Confucius. *The Analects*. Translated by Arthur Waley. New York: Vintage, 1938.

———. *The Book of Rites (Li Ji)*. Translated by James Legge. Oxford: Clarendon Press, 1885.

Connolly, Kate. "AI Preaches at German Church Service." The Guardian, June 9, 2023.

Copernicus, Nicolaus. *On the Revolutions of the Heavenly Spheres*. Translated by Edward Rosen. Reprint, Baltimore: Johns Hopkins University Press, 1992.

Coulson, Charles A. Science and Christian Belief. Cambridge: Cambridge University Press, 1955.

Cox, Harvey. Fire from Heaven: The Rise of Pentecostal Spirituality and the Reshaping of Religion in the Twenty-First Century. Reading, MA: Addison-Wesley, 1995.

Crossan, John Dominic. Jesus: A Revolutionary Biography. San Francisco: HarperSanFrancisco, 1994.

Culver, Denise. "South Korea's Digital Afterlife Industry Uses Holograms to Preserve Ancestors." CNN, November 3, 2022.

Dalley, Stephanie. Myths from Mesopotamia: Creation, the Flood, Gilgamesh, and Others. Oxford: Oxford University Press, 2000.

Damon, Paul E., et al. "Radiocarbon Dating of the Shroud of Turin." Nature 337, no. 6208 (1989): 611–15.

Darwin, Charles. On the Origin of Species by Means of Natural Selection. London: John Murray, 1859.

De Caro, Liberato, Teresa Sibillano, Rocco Lassandro, Cinzia Giannini, and Giulio Fanti. "X-ray Dating of a Turin Shroud's Linen Sample." Heritage 5, no. 2 (2022): 860–70.

Descartes, René. Meditations on First Philosophy. Translated by John Cottingham. Cambridge: Cambridge University Press, 1996.

Durant, Will. The Story of Civilization, Part III: Caesar and Christ. New York: Simon & Schuster, 1944.

Durkheim, Émile. The Elementary Forms of Religious Life. Translated by Karen E. Fields. New York: Free, 1995.

Dyson, George. Turing's Cathedral: The Origins of the Digital Universe. New York: Pantheon, 2012.

Eckhart, Meister. The Complete Mystical Works of Meister Eckhart. Translated by Maurice O'C. Walshe. New York: HarperOne, 2009.

Ehrman, Bart D. Did Jesus Exist? The Historical Argument for Jesus of Nazareth. New York: HarperOne, 2012.

Einstein, Albert. The World as I See It. New York: Philosophical Library, 1934.

Eisenstein, Elizabeth L. The Printing Press as an Agent of Change. Cambridge: Cambridge University Press, 1979.

Eliade, Mircea. The Sacred and the Profane: The Nature of Religion. New York: Harcourt, 1959.

Falls, David. "God's AI Reckoning: The Final Revelation." Interalia Magazine, October 2025. https://www.interaliamag.org/articles/david-falls-gods-ai-reckoning-the-final-revelation/

Fitbit. *Fitbit Tracker*. Fitbit Inc., 2009.

Foucault, Michel. Discipline and Punish: The Birth of the Prison. Translated by Alan Sheridan. New York: Vintage, 1995.

Pasquale, Frank. The Black Box Society: The Secret Algorithms That Control Money and Information. Cambridge, MA: Harvard University Press, 2015.

Frei, Max. "Nine Years of Botanical Research on the Shroud." Shroud Spectrum International, 1982. https://www.shroud.com

Freud, Sigmund. The Interpretation of Dreams. New York: Macmillan, 1913.

———. The Future of an Illusion. New York: W. W. Norton, 1961.

Friston, Karl. "The Free-Energy Principle: A Unified Brain Theory." Nature Reviews Neuroscience 11 (2010): 127–38.

Galilei, Galileo. Discoveries and Opinions of Galileo. Translated by Stillman Drake. New York: Doubleday Anchor, 1957.

Geertz, Clifford. The Interpretation of Cultures. New York: Basic, 1973.

Griffiths, Roland R., et al. "Psilocybin Can Occasion Mystical-Type Experiences Having Substantial and Sustained Personal Meaning and Spiritual Significance." Psychopharmacology 187, no. 3 (2006): 268–83.

Gutiérrez, Gustavo. A Theology of Liberation. Maryknoll, NY: Orbis, 1973.

Harlow, John M. "Recovery from the Passage of an Iron Bar through the Head." Publications of the Massachusetts Medical Society 2 (1868): 327–47.

Heidegger, Martin. Being and Time. Translated by John Macquarrie and Edward Robinson. New York: Harper & Row, 1962.

Heraclitus. Fragments. Translated by T. M. Robinson. Toronto: University of Toronto Press, 1987.

Hesiod. Theogony and Works and Days. Translated by M. L. West. Oxford: Oxford University Press, 1988.

Hoover, Stewart M. Religion in the Media Age. New York: Routledge, 2006.

Hume, David. An Enquiry Concerning Human Understanding. Edited by Tom L. Beauchamp. Oxford: Oxford University Press, 1999.

James, William. The Varieties of Religious Experience: A Study in Human Nature. New York: Longmans, Green, 1902.

———. The Will to Believe and Other Essays in Popular Philosophy. New York: Longmans, Green, 1897.

Josephus, Flavius. *Jewish Antiquities*, Books XVIII–XIX. Translated by Louis Feldman. Cambridge, MA: Harvard University Press, 1965.

Searle, John. "Minds, Brains, and Programs." Behavioral and Brain Sciences 3, no. 3 (1980): 417–57.

Kant, Immanuel. Critique of Practical Reason. Translated and edited by Mary J. Gregor. Cambridge: Cambridge University Press, 1997.

———. Critique of Pure Reason. Translated by Paul Guyer and Allen W. Wood. Cambridge: Cambridge University Press, 1998.

———. Groundwork of the Metaphysics of Morals. Translated and edited by Mary J. Gregor. Cambridge: Cambridge University Press, 1998.

Kierkegaard, Søren. Fear and Trembling. Translated by Alastair Hannay. London: Penguin, 1985.

———. Purity of Heart Is to Will One Thing. Translated by Douglas V. Steere. New York: Harper, 1956.

Kohlbeck, John A., and Eugenia Nitowski. "New Evidence May Explain Image on Shroud of Turin." Biblical Archaeology Review 12, no. 4 (1986): 18–29.

Kübler-Ross, Elisabeth. On Grief and Grieving: Finding the Meaning of Grief Through the Five Stages of Loss. New York: Scribner, 2005.

Küng, Hans. On Being a Christian. New York: Doubleday, 1976.

———. *Theology for the Third Millennium: An Ecumenical View.* New York: Doubleday, 2002.

Leeming, David. The Oxford Companion to World Mythology. Oxford: Oxford University Press, 2005.

———. The World of Myth: An Anthology. Oxford: Oxford University Press, 1990.

Leibniz, Gottfried Wilhelm. Monadology and Other Philosophical Essays. Translated by Paul Schrecker and Anne Martin Schrecker. Indianapolis: Bobbs-Merrill, 1965.

Leong, Mark. "Robot Chaplain Pepper Offers Comfort to Hospital Patients." Reuters, July 12, 2020.

Levinas, Emmanuel. Totality and Infinity: An Essay on Exteriority. Translated by Alphonso Lingis. Pittsburgh: Duquesne University Press, 1969.

Longenecker, Dwight. "Poetry and Prayer in the Age of Artificial Intelligence." Standing on My Head, 2017. https://dwightlongenecker.com

Lucretius. *On the Nature of Things.* Translated by Martin Ferguson Smith. Indianapolis: Hackett, 2001.

Luther, Martin. The Bondage of the Will. Translated by J. I. Packer and O. R. Johnston. London: James Clarke & Co., 1957.

Mackie, J. L. *The Miracle of Theism.* Oxford: Oxford University Press, 1982.

Marcel, Gabriel. *The Mystery of Being, Vol. 1: Reflection and Mystery.* Translated by G. S. Fraser. Chicago: Henry Regnery, 1960.

Marx, Karl. *Critique of Hegel's Philosophy of Right.* Cambridge: Cambridge University Press, 1970.

McCarthy, Simone. "Robot Priest Debuts at Kyoto Temple." *BBC News*, February 23, 2019.

McLuhan, Marshall. *Understanding Media: The Extensions of Man.* New York: McGraw-Hill, 1964.

Meier, John P. *A Marginal Jew: Rethinking the Historical Jesus.* New Haven: Yale University Press, 1991–2016.

Miller, Lisa. *The Awakened Brain: The New Science of Spirituality and Our Quest for an Inspired Life.* New York: Random House, 2021.

Mill, John Stuart. *Utilitarianism.* Edited by Roger Crisp. Oxford: Oxford University Press, 1998.

Mozur, Paul. "Google's AlphaGo Defeats Go Champion Lee Sedol." *New York Times*, March 15, 2016.

Naydler, Jeremy. *Temple of the Cosmos: The Ancient Egyptian Experience of the Sacred.* Rochester, VT: Inner Traditions, 1996.

Nagel, Thomas. "What Is It Like to Be a Bat?" *Journal of Philosophy* 83, no. 4 (1986): 435–50.

Newton, Isaac. *Philosophiæ Naturalis Principia Mathematica.* 1687. Reprint, Berkeley: University of California Press, 1999.

Nickell, Joe. *Looking for a Miracle: Weeping Icons, Relics, Stigmata, Visions & Healing Cures*. Amherst, NY: Prometheus Books, 1993.

Noble, Safiya Umoja. *Algorithms of Oppression: How Search Engines Reinforce Racism*. New York: NYU Press, 2018.

O'Neil, Cathy. *Weapons of Math Destruction: How Big Data Increases Inequality and Threatens Democracy*. New York: Crown, 2016.

Ong, Walter J. *Orality and Literacy: The Technologizing of the Word*. New York: Routledge, 1982.

Oppy, Graham. *Arguing About Gods*. Cambridge: Cambridge University Press, 2006.

Paley, William. *Natural Theology; or, Evidences of the Existence and Attributes of the Deity*. Oxford: Oxford University Press, 2006.

Pascal, Blaise. *Pensées*. Translated by Roger Ariew. Indianapolis: Hackett, 2005.

Pew Research Center. *Religious Landscape Study*. Pew Research Center, 2014. https://www.pewresearch.org/religion/religious-landscape-study/.

Picard, Rosalind W. *Affective Computing*. Cambridge, MA: MIT Press, 1997.

Plantinga, Alvin. *God and Other Minds: A Study of the Rational Justification of Belief in God*. Ithaca, NY: Cornell University Press, 1967.

Plato. *Apology*. Translated by G. M. A. Grube and revised by John M. Cooper. Indianapolis: Hackett, 2000.

———. *Republic*. Translated by Desmond Lee. London: Penguin Classics, 2007.

———. *Timaeus*. Translated by Donald J. Zeyl. Indianapolis: Hackett, 2000.

Polanyi, Michael. *The Tacit Dimension*. Garden City, NY: Doubleday, 1966.

Popol Vuh. Translated by Dennis Tedlock. New York: Simon & Schuster, 1996.

Popper, Karl. *The Logic of Scientific Discovery*. London: Routledge, 1959.

Ricoeur, Paul. *Interpretation Theory: Discourse and the Surplus of Meaning*. Fort Worth: Texas Christian University Press, 1976.

Rig Veda. *Rig Veda: A Metrically Restored Text*. Translated by Stephanie W. Jamison and Joel P. Brereton. Oxford: Oxford University Press, 2014.

Robinson, H. Wheeler. *The Christian Doctrine of God*. Eugene, OR: Wipf & Stock, 2009.

Rosenthal, David M. *Consciousness and Mind*. Oxford: Clarendon Press, 2005.

Russell, Bertrand. "Is There a God?" *Illustrated Magazine* (commissioned but unpublished), 1952. Reprinted in *The Collected Papers of Bertrand Russell*, Vol. 11: *Last Philosophical Testament, 1943–68*, edited by John G. Slater, 543–48. London: Routledge, 1997.

Safron, Adam. *Consciousness and the Brain: Deciphering How the Brain Codes Our Thoughts*. Oxford: Oxford University Press, 2016.

Searle, John. *Minds, Brains, and Science*. Cambridge, MA: Harvard University Press, 1984.

Seth, Anil. *Being You: A New Science of Consciousness*. London: Faber & Faber, 2021.

Shermer, Michael. *The Believing Brain: From Ghosts and Gods to Politics and Conspiracies*. New York: Times Books, 2011.

Spinoza, Baruch. *Ethics*. Translated by Edwin Curley. London: Penguin Classics, 1996.

Swan, Melanie. *Blockchain: Blueprint for a New Economy*. Sebastopol, CA: O'Reilly Media, 2015.

Taylor, Charles. *A Secular Age*. Cambridge, MA: Belknap Press, 2007.

The Holy Bible. New International Version. Grand Rapids, MI: Zondervan, 2011.

The Qur'an. Translated by M. A. S. Abdel Haleem. Oxford: Oxford University Press, 2004.

Tononi, Giulio. *Phi: A Voyage from the Brain to the Soul*. New York: Pantheon, 2012.

Turing, Alan M. "Computing Machinery and Intelligence." *Mind* 59, no. 236 (October 1950): 433–60.

Turner, Victor. *The Ritual Process: Structure and Anti-Structure.* Chicago: Aldine, 1969.

University of Oxford. *Oxford English Dictionary*, online edition. Oxford University Press, 2020. https://www.oed.com.

Watts, Alan. *The Book: On the Taboo Against Knowing Who You Are.* New York: Vintage, 1966.

Weber, Max. *The Sociology of Religion.* Translated by Ephraim Fischoff. Boston: Beacon Press, 1963.

William of Ockham. *Summa Logicae.* Translated by Michael J. Loux. Notre Dame, IN: University of Notre Dame Press, 1974.

Wittgenstein, Ludwig. *Philosophical Investigations.* Translated by G. E. M. Anscombe. Oxford: Blackwell, 1953.

Xunzi. *Xunzi: The Complete Text.* Translated by Eric L. Hutton. Princeton: Princeton University Press, 2014.

Zuboff, Shoshana. *The Age of Surveillance Capitalism: The Fight for a Human Future at the New Frontier of Power.* New York: PublicAffairs, 2019.

www.ingramcontent.com/pod-product-compliance
Lightning Source LLC
Chambersburg PA
CBHW062022220426
43662CB00010B/1438
* 9 7 9 8 3 8 5 2 6 7 9 9 6 *